ISBN: 978-1-925128-32-1

First Kindle Edition: 2015 (Revised in 2023)
First Paperback Edition: 2023

Disclaimer

Dedication

This book is dedicated to all the curious students and readers – senior high school students (and their parents) who are considering Political Science as their field of study, undergraduate students who are struggling to choose their major/minor, and graduate/research/PhD students who want to pursue further studies/research in the area of Political Science.

Table of Contents

Preface

Given the investment in time and money that students will spend on higher education, understanding what they will learn in their chosen major and how it will prepare them for a career upon graduation is very important. As you read this book, we invite you to think about the types of skills that might be helpful for someone pursuing a career in Political Science.

For students considering an undergraduate degree in Political Science, this book has been prepared to assist you in your research of college majors by:

- providing a description of Political Science (which helps you understand what it is),
- identifying courses typically found in this type of degree (which help you focus on what you will learn), and
- discussing career opportunities in the field (which helps you understand why this degree is important for both you and perspective employers).

Additionally, this book will give you information and best practice advice on graduate/PhD studies and scholarships in Political Science. If you are a practitioner/professional in Political Science, you will find the discussion and insights from practitioner perspective interesting and helpful.

Whether you are a freshman/undergraduate student or graduate/PhD student or a practitioner, this book will give your lots of insights and best practice advice concerning the field of Political Science. In 23 Chapters, top professors from prestigious universities have answered answer the questions including:

- Why should you choose Political Science as your major or minor?
- Why should you undertake a graduate/PhD degree in Political Science?
- What are the research areas/issues and scholarship opportunities in field?
- What are the career options and best practice tips for the graduates?
- What are the academic/faculty peer-perspectives as to Political Science education, research and practice?
- What are the key issues and best practices from the practitioner perspective?

A ten-member Advisory and Editorial Board was established as part of the publication/review process of this book. Most of the chapters in this book are peer-reviewed (except short articles and interviews), and they have been written in plain English to give the best/easiest reading experience to the readers.

So, why study Political Science – are you curious to know? You really need to read the book to find the answers but the following paragraphs provide a brief outline of each chapter.

Why should undergraduate students seek a degree in Political Science? This question is addressed by Professors Gordon A. Babst and Lori Cox Han of the Chapman University in **Chapter 1** entitled **Political Science for Undergrads**. Professors Babst and Han note that people who desire change in their communities, their country, or even in the world at large are often drawn to the study of politics. Why? "Politics has this world-changing potential because it concerns the public activity of people who share something in common and have established a public, authoritative framework within which to advance the sort of changes they believe are appropriate given their values, needs, and aspirations" explain Babst and Han. The authors emphasize Political Science as an academic area of inquiry that is subdivided into many areas such as political philosophy and theory, American Politics, public administration, international relations, and comparative politics. As Professors Babst and Han suggest, if you aspire to careers in politics, if you are motivated to seek knowledge in this field, if you like to make a difference in your society...you may be drawn to the Political Science major. In concluding this chapter, Professors Babst and Han emphasize undergraduate research and inform you that funding opportunities are provided through their Office of Undergraduate Research, as well as internships, study abroad program.

"**Why Political Science?: Applying Principles to Practice**" is the title of **Chapter 2** by Dr. Rachel Paine Caufield of Drake University. Why politics is important? "Politics refers to the practice of power and how we exercise influence - it is about the use of force, coercion, and persuasion" explains Professor Caufield. He further adds that the study of political science does not need to be purely academic, nor is it reserved to those who want to be politicians. Professor Caufield says that the field of political science is about the systematic study of the political world and explains the underlying principles including collective action, institutional design, and agenda-setting. Professor Caufield purposes that politics is everywhere around us, deeply embedded in even the most ordinary circumstances of our day-to-day lives, and that political science is an effort to systemically understand the dynamics of social interaction and the processes that produce collective outcomes. The sheer number and range of applications is itself exciting. Professor Caufield concludes this chapter by sharing a learning experience that, in his own life, he has also found that the principles of political science, applied in a practical and intentional way, have empowered him to be a more thoughtful, informed, active, and effective citizen.

"**Showing How the Magician Does A Trick**" is the title of **Chapter 3** by Professor Chapman Rackaway of the Fort Hays State University. As Professor Rackaway suggests, political science is a unique academic discipline because it provides the student an opportunity to pierce the veil of one of the most opaque elements of human behavior: politics. Professor Rackaway wants you to think about any time you have watched a political ad or consumed some political news and if you are curious about the vagaries of human interaction, you will see that there is no better field of study. What are the concentrations within the political science major? Professor Rackaway mentions a few of them in this chapter including Political Management, Political Behavior and Political Parties, Lobbying, Political Communication, Social Media, and Political Campaign Management. Professor Rackaway compares his teaching to Jilette's approach, because he teaches students how the magicians do their tricks. "Knowing how the magician does a trick lets you see the real truth and not the illusion. Since politics is so central to our lives, stripping away the deception is important for the citizen and a special opportunity for the scholar" Professor Rackaway explains. "For the curious mind, the ability to decode –or craft - political messages is a unique element of the study of political science" Professor Rackaway concludes this chapter.

"**Study and Research in the Discipline of Political Science**" is the title of **Chapter 4** by Michael J. Nojeim of the Prairie View A&M University. What are you expected to do as a political scientist? You ask. "Political scientists uncover emerging trends in the media's coverage of politics and government and also study patterns in voting and elections. Experts in a discipline should be able to *predict* what might happen in the future so that government and society can be prepared for what comes next. They must be able to *prescribe* solutions for key domestic and international problems" Professor Nojeim offers his perspectives. What can a person who earns a degree in political science actually do? You ask. "Well, what *can't* they do? Graduates in political science possess skills (such as research, writing, and critical thinking skills) that are applicable to virtually every professional endeavor" Professor Nojeim explains. He also discusses about research issues and makes the point that political science research is most valuable when it helps politicians, government officials and industry officials do their job better. "Indeed, many famous U.S.

politicians, such as Hilary Clinton, Condoleezza Rice and Barack Obama, earned degrees in Political Science. But so too did actor Alec Baldwin, sportscaster Greg Anthony, AOL CEO Steve Case, and Nobel Laureates Ralph Bunche and Elinor Ostrom. Really, the possibilities are endless" Professor Nojeim concludes this chapter.

Political Science: More than Practical Politics is the title of **Chapter 5** by Professor Dinorah Azpuru of the Wichita State University. "When people hear that someone is studying political science they tend to think that they like practical politics—either they want to become politicians or they are interested in making a career running political campaigns" notes Professor Azpuru. You are you one of those who is baffled by the word 'science'. How can politics and science be together? Professor Azpuru asks you. As he asserts you, while it is true that someone who studies political science may be interested in becoming a politician or in becoming a member of a politician's staff, the truth is that the study of politics opens multiple avenues for professional development. Professor Azpuru discusses the four well-defined fields of interest in political science and encourages you to take quantitative courses related to politics. Why? "It will increase your understanding of the world of politics or international relations, as well as your prospects for employment" Professor Azpuru offers his reason. In concluding this chapter, Professor Azpuru talks about the scholarship opportunities makes two suggestions for you: (1) you should be aware that political science is a versatile field of studies that goes well beyond practical politics; and (2) you should not be afraid of numbers as the student of political science.

"**Confessions of an Aging Political Theorist: Apologia Pro Vita Sua**" is the title of **Chapter 6** by Professor Ernest J. Yanarella of the University of Kentucky. How do you find the scope/future of Political Science in terms of career opportunities and options? Why should you choose Political Science as you undergraduate or postgraduate major? Why should you undertake a research degree in Political Science? What are the research areas or priorities in Political Science? Professor Yanarella answers these and other questions in this chapter. As Professor Yanarella suggests, the entering student of politics must never forget that contemporary political science and the inquiry into political ideas and practices remain largely grounded in and defined by the first and enduring words of political theoretical titans who loved theorizing about political things and who sometimes involved themselves in the rough-and-tumble world of politicking. He notes that the political and social world of today is a data-driven and data-facilitated one. Professor Yanarella further acknowledges that political Science can be seen as a profession of the study of the politics and the politics of the study of a profession. "If not Plato, then surely Aristotle would be proud of us for that remembrance and that acknowledgment" concludes Professor Yanarella.

Why Study Political Science? In answering this question in **Chapter 7**, Dr. Bruce P. Lapenson from the North Carolina Central University poses some big questions for discussion: What is government? Why is government necessary? What is freedom? Can someone truly be free living under government? What are rights? What is meant by citizenship? What can government ensure? Does government restrict individual freedom or provide the only possible environment for it? What does it mean to be a citizen? What is the proper role of government and what should it ensure or be responsible for? Does politics influence behavior? Professor Lapenson then turns to the most pragmatic of concerns for you: internship opportunities and career options. In concluding this chapter, Professor Lapenson ascertains that the study of political science has many educational and practical rewards.

Professor Diana Owen of the Georgetown University shares her story as to how she entered the university to pursue a career as an attorney and how she ended up studying political science. In this chapter (**Chapter 8 – Now is an Especially Exciting Time to Study Political Science**), Professor Owen talks about her research which focuses on new media in election campaigns. As she recalls, the 2008 presidential campaign was a laboratory for examining the ways in which voters' innovation with social media helped to reshape electoral communication between the candidates and the citizens who used Facebook, Twitter, Tumblr, YouTube, and other platforms to supply information, express opinions, and create online videos on behalf of candidates and issues they supported. "Every day I find relevance of political happenings to my research and scholarship. I am most fortunate to be able to explore political trends as they unfold with students and colleagues. Now is an especially exciting time to study political science, especially with new phenomena unfolding, such as new media's influence on politics and government" Professor Owen concludes.

Chapter 9 by Professor Dr. Kwame Antwi-Boasiako from Stephen F. Austin State University is entitled as "**Political Science: An Endless Song with a Variety of Rhythms**". In this chapter, Professor Antwi-Boasiako discusses and provides various definitions of political science. "Political ignorance is a disease, which prevents us from actively and positively participating in the political system" argues Professor Antwi-Boasiako. He says that globalization, global politics, foreign policy, public policy, international political economy, international trade, international organization, international sports events, and many other global topics continue to justify the importance of political science. In this chapter, Professor Antwi-Boasiako also explains the benefits of studying political Science at a mid-size university and discusses the future of political science. "Political ignorance may help voters to elect "wrong" people as their leaders...the study of political science equips us in our research, written and oral communication, and helps us to understand the world around us", Professor Antwi-Boasiako concludes this chapter.

Politics: fascinating, factual and fun. "So, you have chosen to study political science or perhaps you are contemplating studying the discipline in the near future? If it is the former, let me be the first to congratulate you on your excellent choice. If the latter is true then, hopefully, this short piece will help to convince you to study the subject that is at the heart of everything we do" so says Dr Jacqueline Briggs of the University of Lincoln in **Chapter 10**. In this chapter, she offers the reasons as to why choosing to study politics is a clever move! "Everything we do involves politics as it concerns power. Having studied and taught politics for more than thirty years, I remain fascinated by why some people are passionate about politics and why others have not yet caught the 'bug'" Dr Briggs shares her learning experience with you. What can you expect to learn? As Dr Briggs says, heated classroom debates may mean that you have to 'agree to disagree including on brain-aching topics including: Is human nature basically good or basically bad? Do we need lots of rules, laws and regulations almost to protect us from ourselves? Should capital punishment be allowed, should a state be able to take life as the ultimate form of punishment? Should gay marriage be allowed? Should sixteen year olds be able to vote? "If you haven't already made the decision to study political science as your major, hopefully, some of the comments above will enable you to make the right choice. Believe me, you won't regret it!" Professor Briggs wants to assure you in concluding this chapter.

Dr. Lowell W. Barrington has contributed Chapter 11 entitled "**Political Science and Its Study at Marquette University**". In this chapter, Dr Barrington answers a number of questions to satisfy your curiosity. What is Political Science? What are the career opportunities like for Political Science majors? How about scholarships and resources for students? "Because politics touches on so many aspects of economic activity, social relations, and daily life, developing the ability to understand it opens doors to careers in the wide-ranging areas of government, law, business, advocacy, and service just to name a few. We often say that it is easier to list the jobs you *cannot* do with a political science degree than all of the ones you can" Dr Barrington asserts you. She also shares the good news for you in this chapter: "There are also many scholarship opportunities for undergraduate and graduate students from other countries. Marquette University works hard to provide as much financial support to its domestic and international students as possible. Historically, 90% of our international students have received scholarship support". Dr Barrington talks about the practitioners/research issues in Political Science and concludes this chapter with some tips for your career success: The best political science students – and the ones who are the most successful at whatever they choose to do after college – are those who listen to, take seriously, and really try to understand the arguments of people with whom they disagree.

"It is no accident that Aristotle, the great Greek philosopher, described Politics as the master science" notes Professor Robert Garner of the University of Leicester in **Chapter 12** entitled **Politics is the Master of Science**. It is preeminent because politics is an inevitable feature of all societies. "We cannot do without politics, it is inevitable" declares Professor Garner. Why? Politics is inevitable because of three features of the human condition. Professor Garner explains the three features of the human condition and offers his reasons to satisfy your curiosity as to why study politics in this chapter. He talks about the academic research areas in politics that can have a considerable societal and economic impact. "Interest groups and political parties are always on the look-out for research findings that they can utilise to further their cause" Professor Garner concludes this chapter.

Chapter 13 by Professor Thomas Sutton of Baldwin Wallace University is entitled as "**Why Study Political Science**?" As Professor Sutton notes, in an era of constant pressure to go to university, to prepare for a career in a field with good hiring prospects, to earn enough income to pay off student loans, buy a car, buy a house, and live comfortably in an expensive society, it is little wonder that political science can appear to be a 'questionable' choice as a field of study for undergraduate students. Against this background, Professor Sutton wants to help students understand why political science is a good field of study as an endeavor in intellectual development, improvement of necessary career-related skills, and an enjoyable pursuit in and of itself. What do students learn to do in Political Science? How to conduct empirical research and analysis? What study abroad and internship opportunities are available? Professor Sutton answers these questions for you and concludes this chapter with these words of assurance: Rest assured, if you grasp the answer to 'why study political science?', the question of 'what do I do with a political science degree?' will be answered by your own interests, skill development and ability to seize opportunities as they arise.

"**Political Science is to do Beyond Political Science**", so says Dr Hyun Yun of the Texas State University in **Chapter 14**. Dr Yun offers her reasons: In a practical and direct perspective, if you want to work for the government, public agencies, international organizations, or non-profit organizations, studies in political science serve as the required qualification. Dr Yun lists a number of more tangible benefit of studying political science in this chapter and further adds that the study of political science educates you not only to be a well-trained political researcher and practitioner in the field of politics, but also enables you find your real self and understand others under the given political and historic circumstances, and lets you live with others, adjusting to the changes in human history as a valued world citizen. Dr Yun asserts you that political science teaches you the past, current, and future trends and lets you float with the timely political, social, and economic waves. "Without understanding the chronological changes in trends, you cannot appropriately work with others and cannot be on the same page with others in the current time" Dr Yun concludes this chapter.

Why Study Political Science? In **Chapter 15**, Professor Jeb Barnes from the University of Southern California provides **the answers for Homo Politicus and Homo Civicus**. "This question is deceptively complex" notes Professor Barnes. "To answer it, we must first define political science and then consider the question from the perspective of two very different types of people, what Robert Dahl once called "homo politicus"—individuals who are naturally interested in politics or self-proclaimed "political junkies"—and "homo civicus"—individuals who are responsible citizens but do not have any burning interest in the minutiae of politics or policy-making processes" Professor Barnes explains. He notes that media often fail to gives us any insight into the crucial questions of *why?* and *so what?* It is for this reason Professor Barnes makes the point that, to make a difference on issues that you care about, you must be able to identify policy options and read the institutional landscape to see when and where there are chances to act; political science offers a powerful lens for seeing the range of policy options and the institutional opportunities for action. As Professor Barnes concludes this chapter, political science embraces diverse strategies for studying complexity and studying these strategies in depth can give students tools for understanding the world around them.

Would democracy grow or fail globally? In **Chapter 16 (Why Study Political Science?)**, Professors Greg Flynn and Ahmed Huque of McMaster University attempt to define political science by looking at the literature.

Professors Flynn and Huque discuss the various concentrations within the discipline including international political economy, international relations, public administration and many more tracks in political science. Professors Flynn and Huque hope that the various definitions presented in this chapter should encourage you as a student to be able to define political science the way you understand it. In this chapter, Professors Flynn and Huque also looks at as to why you should study political science from the career/practitioner perspectives, and what are the benefits of studying political science at a mid-sized university. "The general consensus is that political ignorance may help voters to elect "wrong" people as their leaders. Hence, the study of political science equips us in our research, written and oral communication, and helps us to understand the world around us" Professors Flynn and Huque conclude this chapter.

Professor Zehra Arat of the University of Connecticut has contributed **Chapter 17** entitled "**Value of Political Science for Political Participation**". Professor Arat argues that students of "politics" have conceptualized the term differently, but a common broad understanding of it can be simply summarized as "struggle for power." Why? Professor Arat starts with two important arguments related to this particular definition of politics: First, struggle for power takes place at *different levels* and at *different sites*. Second, *politics* are not about a set of events, such as elections, parliamentary debates or legislation, but should be understood as an ongoing complex *process* that involves multiple events and processes. In order to justify that struggle for power is a continuous process, Professor Arat quotes Karl Marx: "Philosophers have hitherto only *interpreted* the world in various ways; the point is to *change* it". Professor Arat then makes the point that, if we attempt to *change* the world – or simply our lives – we would need to be engaged in politics *consciously*. "If we want change, we need to be cognizant and take action. No action *may* mean maintaining the status quo" Professor Arat explains. In concluding this chapter, Professor Arat makes a distinction between the two types of politics – high politics and everyday politics and asks you this critical question: "Do you want to be engaged in politics incidentally or consciously, by being aware of the consequences of your action and non-action?"

Why Study the Politics of Human Rights? Professor Clair Apodaca from Virginia Tech explores this question in **Chapter 18**. Are you interested in human rights issues? "Students who study the politics of human rights are interested in the nature of contemporary human rights problems and the role that politics and political regimes play in generating them, the economic, social, and political factors associated with human rights progress, and the strategies currently being employed to improve human rights in different settings" says Professor Apodaca. She poses some questions for you: Why do some states commit more human rights violations than others? What role does political order and regimes play in state repression? How do power inequalities, discrimination, and poverty shape a state's commitment to combat human rights violations? How can one evaluate the implementation of human rights obligations in practice? Professor Apodaca discusses these questions and outlines several course related to human rights including women's rights, global human rights issues, human rights issues and perspectives, global poverty and welfare, Indigenous peoples and human rights, and civil and political rights. In concluding this chapter, Professor Apodaca asserts you that human rights are an important force in global, regional, and domestic politics and will remain so for many years to come.

Professor David A. Dulio of the Oakland University has contributed **Chapter 19** entitled "**Students should Choose a Major in Political Science for Several Reasons**". What actually is the academic discipline/major of Political Science? Professor Dulio believes you should choose a major in political science for several reasons: First, the content you will study in any area of political science will be important for you as you engage as a member of our global society. Moreover, and possibly more importantly, the skills that you will learn in political science courses are skills that can be used in just about any career or occupation. Why undertake a research degree in Political Science? "One of the great things about studying political science is that there are so many questions yet to be asked or addressed...When you undertake a research degree in political science, you can be a part of the world of making great discoveries about questions and phenomena that interest you" explains Professor Dulio. 'The findings from research projects can inform the work the political practitioners do and even help them do their work" Professor Dulio concludes this chapter.

As discussed in the chapters above, Political Science is an interdisciplinary field of with many sub-disciplines. For this reason, we have included chapters from the books in international relations, public policy, philosophy, and public speaking. For details of books, you should read the books from the Curious Academic Publishing.

In "**Why I Love Studying International Relations**?" (**Chapter 20**), Professor Kenneth A. Schultz from Stanford University shares his personal childhood story to make his case and offers reasons why he loves studying international relations. Professor Schultz poses some interesting questions: Why are there wars? Why are some people and nations poor while others are rich? Why do some live in freedom, while others live in tyranny? Why do people sometimes take up arms against their governments or foreign states? Why is it so hard for people and countries to cooperate to stop genocides, resolve humanitarian crises, or prevent catastrophic environmental damage? Professor Schultz maintains that, as students of international politics, we learn from these kinds of events and develop tools to understand them. In this Chapter, Professor Schultz discusses these questions and suggests that game theory provides a way to think systematically about strategic situations, in which the outcome depends on the choices of multiple actors with different goals. Professor Schultz suggests us to "think strategically" with an example of a foreign policy "mistake" which resulted from a failure to think strategically. According to Professor Schultz, failure to correctly anticipate the actions of others, or to interpret the signals they try to send you, can lead to bad outcomes. His advice to students and researchers?: "Analyze What Went Wrong - Don't Make the Same Mistakes". What is likely to be the most lasting impact of his career? "There is little question in my mind that it is through the students that I have taught", says Professor Schultz. For details, please read the book *International Relations for the Curious* published by the Curious Academic Publishing.

What's Studying Philosophy Good for? Professor Hans-Herbert Kögler from the University of North Florida asks you in **Chapter 21**. In this chapter, Professor Kögler claims that the beneficial nature of philosophy can only be fully understood if we see how philosophical thought and practice relate to cultural traditions as well as social interaction and institutions. He answers a number of questions to satisfy your curiosity: what is the lifeblood of philosophy and why should you embark on the study of philosophy? What is it that makes this endeavor a great investment in your life's prospects? What is learned and conveyed by its study that both explains the subsequent success in 'real life' of those with degrees in philosophy, and equally defines the deep excitement and satisfaction that every true student of philosophy readily testifies to? Professor Kögler answers these questions and discusses a number of issues including the impact of philosophy, critical thinking, philosophy as reflexive practice, and philosophical dialogue as political ideal. In closing this chapter, Professor Kögler provides the four reasons for studying philosophy. "That's what studying philosophy is good for. Now who is still surprised that philosophy students land on their feet when looking for a job?" Professor Kögler concludes this chapter. If you want to read more about this field of study, you should read the *book* *Philosophy for the Curious*.

In this chapter (**Chapter 22 - Why Should Students Choose Public Speaking as their Undergraduate or Graduate Major/Minor**?), Professor Joseph Burns of the Southeastern Louisiana University argues that Public Speaking, like singing, playing the drums and playing the violin is a skill and that none of us are born with the ability to do it well. "It truly is not what you say but how you say it. It is a skill and it must be learned, practiced and perfected" explains Professor Burns. He notes that at just about any job, in just about any field, employers are screaming for new employees with "communication skills." But think about that - "how do you prove to an employee that you have communication skills?" "The easiest way I know is to have a major or a minor in Public Speaking. That way you can show a possible employer your skill set and what you are able to do for his or her company communication wise" Professor Burns suggests. Is Public Speaking for you? Professor Burns further suggests that in any occupation where you use your voice as your main tool, a degree in Public Speaking is a strong choice. "Have you heard the statement that statistics shows that people are more afraid of Public Speaking than death?" Professor Burns asks you. "Think back. ...The first time you were with your boyfriend or girlfriend. You were frightened then as well but you got past it. You overcame the fear and what you do has become second nature. It's the same with Public Speaking" Professor Burns assures you. "As I said at the beginning of this piece, everyone can talk. Not everyone can "speak." Professor Burns concludes this chapter.

Again, **why study Political Science**? Professor John Dunn from the University of Cambridge offers the reasons in **Chapter 23**: "The best reason to study politics is that it will be what decides the kind of future human beings are going to have for many centuries to come. To study it you must study why human beings find it so hard to live graciously and generously with each other beyond their own families, the arrangements they have so far thought of to handle that task for them, and why those arrangements have worked so poorly". Professor Dunn believes that, over time, politics has done us far more conspicuous harm than clear good. He has written three books over the last twenty five years which try to define this challenge. Professor Dunn wants you to understand that you will think about the challenge much better, and have a far better chance of seeing how to tilt that balance back in a less grim direction, if you spend some of your time while you are still relatively free in learning what really is going on in politics and why it is happening like that. "That is the opportunity which studying political science should offer you. The political scientists of the world must make sure, if you choose to take it up, that that is the opportunity you get" Professor Dunn concludes this chapter.

Chapter 24 is both for teachers and students of Political Science. In this chapter (**Using Humor To Help Students Get Comfortable With Political Science And With You, The Professor**), Professor Thomas Sutton, Ph.D. of Baldwin Wallace University, Ohio, discusses how humor can help the professors. Professor Sutton argues that using humor as a teaching technique helps to get students' attention and to develop a sense of connection between the professor and the class, which helps builds students' interest in the course content. "Using humorous video clips, political cartoons, goofy jokes, and sharing personal stories are all ways in which you can build these connections and help students learn", adds Professor Sutton. The best part? "Teaching becomes a lot more fun for you!", concludes Professor Sutton. If you are interested in learning more about teaching Political Science with humor, you should read the book "*Teach Political Science With a Sense of Humor: Why (and How to) Be a Funnier and More Effective Political Science Educator and Laugh All the Way to Your Classroom*" published by the Curious Academic Publishing.

Now about this book itself. This is the first book of its kind ever published on Political Science education, research and practice. In this book, we have invited a number of professors from top universities to share their perspectives and advice on Political Science to help students make an informed decision about their major/career.

At the Curious Academic Publishing, we have worked hard to bring the books of your interest in plain English. Please search your intended/current discipline on Google or Amazon for other books from the Curious Academic Publishing.

If you purchased this book through an online retailer such as Amazon and liked it, please leave an honest review.

We, at the Curious Academic Publishing, hope you benefit from this book. Whether it's about making an informed decision about your major/minor or career options, or keeping yourself updated about the research trends/scholarship opportunities, this book will prove to be a valuable resource.

We wish you happy reading and all the best with your studies, research, and career in Political Science.

K. Vaidya, PhD
Editor-in-Chief
The Curious Academic Publishing
www.Amazon.com/author/college

Chapter 1: Political Science for Undergrads – Professor Gordon A. Babst, PhD. and Professor Lori Cox Han, PhD

Gordon A. Babst, PhD
Associate Professor of Political Science
Chapman University

Dr. Gordon Babst *is Associate Professor of Political Science, and first joined the faculty at Chapman University in 1999. He received his Ph.D. from the Claremont Graduate School. His major area of emphasis is political philosophy, and his research and teaching interests include religion and politics, theory and globalization, and issues of diversity and tolerance in the liberal-democratic political regime. Dr. Babst is currently working on a manuscript under contract with SPRINGER on Kantian international political theory to be published in the SPRINGER series: Studies in Global Justice, series editor Deen Chatterjee.*

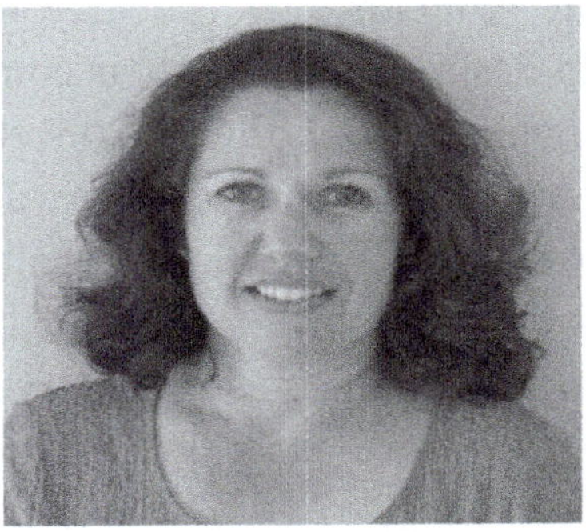

Lori Cox Han, PhD
Professor of Political Science
Chapman University

Lori Cox Han is Professor of Political Science at Chapman University in Orange, California. With research and teaching interests in the presidency, women and politics, media and politics, and political leadership, she is the author of several books, including In It to Win: Electing Madam President (forthcoming in 2015, Bloomsbury Publishing), Presidents and the American Presidency (Oxford University Press); A Presidency Upstaged: The Public Leadership of George H.W. Bush (Texas A&M University Press); Women and U.S. Politics: The Spectrum of Leadership, 2nd ed. (Lynne Rienner); and Governing From Center Stage: White House Communication Strategies During the Television Age of Politics (Hampton Press). She is past president of the Presidency Research Group, an organized section of the American Political Science Association devoted to the study of the presidency. She received her Ph.D. in Political Science from the University of Southern California.

Why should undergraduate students seek a degree in Political Science?

Why should undergraduate students seek a degree in Political Science? In a broader sense, people who desire change in their communities, their country, or even in the world at large are often drawn to the study of politics. Through critical political engagement, people believe they will be able to address the injustices they believe to exist, rectify inefficient or illegitimate uses of power, and resolve divisive issues that can only be addressed through political action. Good government is but a tool of the society it serves, and public service or elected office has long been regarded as a noble calling and as a way to realize our ideals.

As such, the study of politics interests people because they believe that through the careful analysis of human political behavior in all its many manifestations, they may be able to craft interventions into the way things are to make them better for their society, or even humanity in general. Politics has this world-changing potential because it concerns the public activity of people who share something in common and have established a public, authoritative framework within which to advance the sort of changes they believe are appropriate given their values, needs, and aspirations. Wherever people exist, there is a politics to how they relate to one another because the inherent diversity of perspectives, moral outlooks, and goals people have demand coordination and principled justification to those who are affected by any use of public power. Through the critical study of the practices of government, institutions, laws, and regulations, political scientists assess how well a society measures up to its political principles, and whether the exercise of the power to govern is being used appropriately and wisely.

These issues provide a common theme in how we approach the study of political science at the undergraduate level. Specifically, we emphasize Political Science as an academic area of inquiry that studies human political behavior through a variety of methodologies both quantitative and qualitative in nature. Most people develop an intuitive awareness of the impact of politics on their lives, and many have direct experience of government at some level, such as through law enforcement or public policies that benefit them or their families. Because while growing up most people have experiences with governing institutions, and also because of the political ideals they have already formed or were encouraged to accept, Political Science is an attractive undergraduate major as it may seem to be about a field of study they already know something about. Soon they will discover that their ordinary understanding and casual conversations about political issues may contain great depths worthy of continued serious reflection and study. They will also discover that the discipline is subdivided into many areas such as the five subfields we teach here at Chapman University: political philosophy and theory, American Politics, public administration, international relations, and comparative politics.

Political Philosophy or Theory

Students interested in political philosophy or theory tend to be attracted to the study of ideas, concepts, and the political thought of thinkers such as Aristotle,

Machiavelli, John Locke, and, more recently, Hannah Arendt and John Rawls. These thinkers were concerned with justifying the use of coercive power to achieve a peaceful, orderly society in which people can live their lives as freely as possible. Rulers who govern arbitrarily and abuse the freedoms and rights of the people to whom they cannot be held accountable are, sadly, a common phenomenon, both historically and in the world today. Tyrants of whatever stripe who use the public power at their disposal to inflict injury on the people or to maintain a status-quo that works only to the advantage of the few, pose a threat to people's well-being, yet often people accept such abuses believing they have no other choice.

Students interested in American politics study the various institutions and political processes at the national, state, and local levels. Specific topics include the cultural and ideological background of the American system of government; the U.S. Constitution and federalism; civil rights and civil liberties; voting and public opinion; political parties and interest groups; the mass media; campaigns and elections; institutions of government; and the policymaking process. The goal is to help students develop not only critical thinking skills about American government and politics, but to learn how to apply both theoretical and practical knowledge about the political process in becoming more informed citizens. Because the U.S. Constitution has been emulated worldwide, and because of the stability of the American form of government (a democratic republic), it is worthy of serious study even though over time we have realized its shortcomings in serving minorities and in securing the full meaning and value of the freedoms it stands for.

Closely related to American politics is the study of public administration. While as a topic it may seem unexciting at first blush, its importance cannot be overemphasized because it is through sound administrative practices that reflect accountability, transparency, and competence, that our ambitions to effect positive change in important policy areas such as the environment, education, and health care, can be realized.

What are International Relations and Comparative Politics?

The study of international relations has a built-in allure for students because it looks at political phenomena beyond the level of the nation-state, seeking to learn what characterizes relations between countries, within regions, and across the globe. Of course, this subfield seeks to understand compelling political phenomena such as war and other forms of interstate conflict, as well as the nature of the international system and its role in conditioning local politics.

Finally, the subfield of comparative politics looks into societies whose political evolution may have taken a different trajectory than ours, or whose practices of government may provide us with insights into how we may better use power to serve the interests of the people and realize their ideals. Issues of distributive justice that affect the unequal development of different countries and limit the aspirations of their peoples, and comparing the effectiveness of governing institutions and practices, are some of the concerns here.

In addition to a broad range of courses that cover the traditional subfields within the discipline, students who study Political Science also benefit from doing research because it helps them to understand politics better, and to think more critically about the political issues and concepts that interest them the most. Learning how to conduct research in Political Science also prepares the student for advanced study in a graduate degree program, or for success in careers such as journalism, law, government, education, and business, among others.. Research too increases the body of knowledge available to us all, and enhances our understanding of the human condition and better ways to navigate its known pitfalls.

Prospective students may well be drawn to the Political Science major because it appeals to those motivated to seek knowledge in the discipline, or who aspire to careers in politics, or who might like someday to make a difference in their society based on the knowledge and understanding they have acquired. Regardless of the student's interest, he or she will find a faculty member actively engaged in research and the opportunity to participate in that research endeavor in the Political Science department at Chapman University.

What Funding Opportunities are Available?

We strongly believe that training students to conduct research both benefits their development as political scientists, and complements our teaching. We bring our research interests into the classroom, as reflected in our curriculum which features several courses designed around a faculty member's area of expertise. Our research agendas inform our teaching, and, our teaching and interactions with students also help to inform our research agendas.

As such, because we emphasize undergraduate research alongside or under the guidance of a faculty mentor, Chapman University provides funding opportunities through our Office of Undergraduate Research, as well as internships, study abroad, and our semester in Washington, DC program. Through our Alpha Beta Chi Chapter of Pi Sigma Alpha, the national Political Science Honors Society, students acquire leadership skills and an opportunity to network with others pursuing similar interests in the study and practice of politics.

Relevance of Academic Research and Analytical Tools

Alas, it is often the case that political leaders do not have the benefit of an education in Political Science, and refrain from seeking out the advice of those who do. A political explanation for this might be that to take advantage of academic research in Political Science, especially widely accepted empirical studies, would compromise their aims, expose their motivations, or simply not serve the particular constituency they rely on to get elected. In the United States this may reflect ordinary Americans' attitudes toward expertise, or their arguably democratic inclinations not to be lectured to by people with advanced degrees. The obvious downside of this negligence of solid research findings is that political leaders will fail to craft that policy and law implicitly or even emphatically suggested by the research, potentially prolonging arrival at actual solutions to vexing issues of importance to the people.

While it is rare that research leads to only one policy outcome, it is foreseeable that political actions taken that are not grounded in the analysis political scientists can contribute is likely to lead to less than optimal outcomes. This result is all the worse when its burdens must be shouldered by society's most vulnerable, whose voices are likely less heard, and whose interests are less represented.

The real-world impact of political engagement, informed by the careful study and tools of analysis the student learns in the Political Science major, can be quite meaningful, and we strive to impress on students this aspect of the discipline. Hence, we incorporate actual events and controversies, both historical and contemporary, into our teaching, to drive home the point that there is indeed a point to the study of politics, but this is one each student must feel free to discover on his or her own and relative to any political aspirations he or she may have. We do not prescribe any particular understanding of politics to our students, but illuminate a variety of possible approaches for them to arrive at their own understanding of what ought to be, and how political action can get us closer to that.

Chapter 2: Why Political Science? Applying Principles to Practice – Professor Rachel Caufield, PhD

Rachel Paine Caufield, PhD
Associate Professor
Drake University Department of Political Science

Rachel Paine Caufield teaches in the Department of Politics and International Relations and is the Associate Director of the Harkin Institute for Public Policy and Citizen Engagement at Drake University. She received her Ph.D. in Political Science at The George Washington University and her B.A. in Mathematics and Political Science from Hood College in Frederick, MD. Her teaching and research focus on American political institutions, including judicial politics, the American presidency, presidential nominating processes, and legislative politics. For the past two election cycles, she has been the Director of the Iowa Caucus Project at Drake, organizing candidate visits and campus participation in the caucus process and serving as the lead coordinator for the ABC Republican presidential candidate debate on campus. She spent a decade (2003-2013) as a Research Fellow for the Hunter Center for Judicial Selection at the American Judicature Society. Prior to her arrival at Drake, she was a visiting Research Fellow at the Brookings Institution in Washington, D.C.

Why Some People Get What they Want While Others do not

We all live in and among institutionalized social systems. We participate in teams and organize clubs. We go to schools. We belong to families. We are hired by workplaces. We shop in stores. We eat food. We watch TV. Every day, whether we are conscious of it or not, we interact with dozens of social structures that organize our lives. Each of those social structures is a site for politics.

When we say "politics," we often mistakenly assume that it refers exclusively to formal governmental institutions or systems of governance. Politics is certainly present in local and national elections, in policy-making by legislative bodies, in meetings and deliberations of the United Nations. But politics is larger than nation states and elected officials. "Politics" refers to the practice of power and how we exercise influence. Politics is about the use of force, coercion, and persuasion. It helps us understand why some people get what they want while others do not, how we think or talk about a situation or event, what choices we make regarding our schools, our shops, our homes, our food, our jobs, our families, and every other venue of individual action. Every day, when we make decisions and interact with those around us, we are encountering "politics."

The field of political science is about the systematic study of the political world. It includes everything from the most esteemed hallways of power to the mundane experiences of every citizen. Our lives are lived based on expectations about patterns of behavior, institutional dynamics, social constructs, and a range of individual interests that all come into play in virtually all aspects of the public and private spheres.

For example, consider a school. Students may have little familiarity with the wide range of political decisions that define their education. Rules govern who goes to school and what schools they will attend as well as how much money will be dedicated to support public schools. Politicians and school administrators create ways to measure educational outcomes – usually in the form of testing – that determine the students' futures. Politics determines who can be a teacher and how much teachers will be paid, what text books will be used, what subjects will be taught, how many days of the year schools will hold classes, the condition of the school buildings, and what behavior is allowed or punished. There is no "given" school. Educational systems are created and maintained through a series of choices, each of which is the result of the people involved. Each of those people has a set of interests – things that they hope to achieve – and works within an institutional environment that defines how the decision will be reached. And each of those institutions is created by political decisions.

Just as schools do not naturally exist absent political decisions, change does not naturally happen absent political decisions. If we want to change our school systems, we must recognize and understand the complex web of political dynamics that define how the schools operate and what change is possible. If we want to change any aspect of our lives, we have to first understand the political principles that support and maintain the system.

How does Politics Help us?

As another example, in 2011 an earthquake and tsunami resulted in a massive nuclear crisis at the Fukushima Daiichi facility in Japan. In the wake of the disaster, many asked questions about how such a crisis came to be. The answer is a complicated and interactive understanding of why nations choose nuclear power as a source of energy, why this site was chosen for a nuclear power plant, how regulators evaluate safety concerns, how governments work with private businesses in achieving public goals, among many other considerations. The earthquake and tsunami may have been natural disasters, unforeseeable and unpredictable. But the resulting nuclear disaster was not *natural* – it was the result of humans acting within a political system (both domestically and internationally) before and during the crisis. If we seek to avoid another catastrophe, our first job is to understand the range of actions that led to the 2011 crisis and how they reflect the system in which those decisions were made.

Political scientists study a range of topics, most commonly grouped under the broad subfields of American Politics, Comparative Politics, International Relations, Political Theory, and Public Policy. Although each subfield focuses attention on different questions, they all have common understandings of a few fundamental questions or concepts, each of which is essential to evaluating and assessing the vast array of political forces that affect us every day. Because politics manifests itself in all social groupings, the study of these principles not only helps make sense of nation-states, social movements, government institutions, bureaucracies, international organizations, and so-called interest groups. The same principles apply to virtually any social setting.

In other words, the study of political science does not need to be purely academic, nor is it reserved to those who want to be politicians. As a discipline, political science offers ways of thinking that will apply equally well to all aspects of our lives. It helps us diagnose problems in decision-making processes. It helps us negotiate our role as citizens, finding ways to influence government and organizations. It helps us take practical steps to analyze our own behavior relative to our friends, families, co-workers and policy-makers in an effort to be effective advocates and agents of change.

What are those Principles?

Collective Action

At its most basic level, the concept of collective action refers to the ways that groups of people overcome barriers to work together toward some agreed-upon goal. Even in the most basic scenarios, working within any group of people is hard. Anytime you try to work in a group, you will encounter disagreement because each member of the group has their own ideas of the best way to get things done. Finding a way to resolve those differences requires a good deal of time and energy. The theory of collective action starts with the premise that every member of the group has a set of individual interests. In small groups, people know each other and the range of interests is usually fairly small, so disagreement may be easily resolved because interpersonal relationships help us negotiate and compromise so that the group's collective goals can be realized. But when groups start getting bigger, these difficulties increase exponentially, and the reliance on good will among the members will be insufficient. Intuitively, we know this. Working with a group of 3 close friends to get something done is easier than working with 150 people, most of whom we don't know well. In the same way, dynamics among our immediate nuclear family are very different than those within our extended family. The U.N. General Assembly represents 193 nations. How do those delegates find common ground? As citizens, we may live among millions of other citizens. How can the citizenry negotiate disagreement to arrive at a single decision?

Studying the principles of collective action, we can predict problems and assess which strategies are most likely to be effective. To illustrate, consider an example:

Imagine that we live in a great neighborhood with a river running through the middle. About 500 families live in our neighborhood. The river, however, is polluted. A company has moved into an adjacent town upstream and the pollution from their factory is running into the river. Over time, the pollution has killed off much of the wildlife and the river is now causing health problems for the children and pets in the neighborhood. What can we do?

Our first answer may be to gather our neighbors together and encourage them to petition the company to stop polluting the river. As a group, we may also hope to put pressure on our elected representatives to change the rules and force the company to stop. But first we have to form a group. And doing so will prove problematic for a few reasons, all of which are derived from the theories of collective action.

First, we have to get our neighbors to agree to join our group. This will require substantial time and energy on our part to reach out to people and convince them to join us. Some neighbors are more affected by the polluted water than others, so we may be unable to agree, as a community, that this is a priority. How can we convince those who live farther away from the river to be active contributors to our campaign to stop the pollution? Even more problematic is the fact that, if we are successful, all of our neighbors will benefit. If they can benefit without ever contributing any effort to the group, how can we persuade them to join? It's less time-consuming, cheaper, and easier to realize the benefits while doing nothing, so why would they join the group? We may seek to partner with communities even farther downstream, neighborhoods that would be experiencing similar difficulties with the water. But we will encounter the same problems in those neighborhoods. Finally, given that the company's factory is located in another town, the local officials who govern that geographic area are not the elected officials who are charged with representing our neighborhood. They likely see significant benefits to having the company in their town, and since the pollution will disproportionately affect those in other communities, they may not see a reason to stop the company from polluting the river.

While this example may seem contrived, it illustrates a number of challenges (coordination, the tragedy of the commons, the free rider problem, externalities) that are implicated in situations that occur everyday around the world. Why do some countries sign onto international treaties while others do not? Why are some efforts to increase food safety successful while others are not? Why have some towns successfully developed thriving business communities while others have trouble maintaining the businesses they have? Why are some industries unionized and others are not? How are some countries able to take advantage of natural resources while others remain poverty-stricken despite significant natural resources?

Underlying each of these situations, collective action can help us understand why some collective goals are achieved and others are abandoned or left unfulfilled.

Institutional design

One potential solution to collective action dilemmas is institutional design, which refers to the rules and processes put into place to channel and shape decision-making among a group of people. How institutions and the mechanisms of decision-making within institutions are crafted can alleviate or exacerbate common collective action problems. Questions of institutional design are addressed by constitutions and written law, but they do not need to be written in order to be important or recognized.

For example, among friends you may have a custom or tradition of deferring to one person when making a decision about what the group will do on any given weekend. This custom or tradition is a form of institutional design; you could vote on proposals from each person in the group, you could take turns deciding what to do, you could always do the first thing that is suggested. Whether it's a conscious choice or not, the process of deferring to one individual is a way of arriving at a decision and resolving a collective action problem. Informal processes, when followed regularly, can evolve into something resembling institutional design.

Institutional design matters. To return to our neighborhood example, we face a problem because elected officials in the town where the company is located will be unconcerned about our problems of pollution downstream. In any system where representatives are chosen to express the concerns and protect the interests of geographically defined towns and communities, those who are elected will *only* have a vested interest in the concerns of citizens in the immediate locality. Our group, therefore, may easily get the attention of our own representatives but is unlikely to find support from officials elected in the town where the factory is located. While our elected officials have an incentive to listen to us, those elected elsewhere have no incentive to respond to those of us living downstream in another jurisdiction. The system of electing officials to specific geographic areas is a disadvantage for our group's efforts to eliminate the pollution the company creates. If representatives were elected in some other way – i.e. if the institutional design were different – that problem may not arise.

Across a wide range of settings, we see unique institutional features that help us understand the outcomes that we see. A dictator needn't depend on the citizens for their authority and therefore, we should expect that they will make different decisions than leaders in democracies who do depend on citizen support. Parliamentary democracies are different than presidential democracies in predictable ways because the institutional rules are different, creating different patterns of accountability and responsiveness. The U.N. Security Council allows five powerful nations to exercise a veto, which changes the way decisions are made.

Anytime we channel decision-making through institutions with established procedures, the structure of the institution will empower some and disadvantage others. Along the way, the structure of decision-making changes how the wishes of the group (be it a group of friends, a family, a legislature, an interest group, a nation, or an intergovernmental organization like the United Nations) are translated into a specific outcome.

Agenda-setting

Finally, issues, problems, policy recommendations, and priorities rise to the public consciousness in predictable and patterned ways. Agenda-setting refers to the ways that specific concerns are brought to prominence and addressed by the community at large. Although events may call attention to issues, the public agenda – the collection of current topics that are under discussion – is defined through a complex process involving events, organizations (like the media), individuals, and institutions. Decisions are made only when we, as a community, are aware of a "problem" and perceive that there is some achievable "solution." Although we may think that issues emerge organically and proposals to solve problems arise naturally, the process can be hastened or thwarted by those who understand how to set an agenda.

Every day, we confront situations where agenda-setting matters. Think of the parent who quiets a crying child by suggesting the family go for ice cream. Essentially, the parent is simply changing the conversation. Rather than focusing on the thing that the child is crying about, the parent suggests a new activity that will please the child. The new activity may have nothing to do with original reason for the child's displeasure, but shifting attention to a new set of issues (i.e. ice cream) is an effective way to resolve a situation. Instead of suggesting ice cream, the parent might also change the agenda by suggesting the child make a choice between taking a walk, playing with a toy, or eating a carrot. Mom or dad has effectively changed the conversation, but has also shaped the range of alternatives available. The kid may prefer a fourth option (say, playing fetch with the family dog), but that option has not been made available. Just as a parent can change the agenda, social groups focus on some issues over others and we may be constrained in our choices by those with the power to set the agenda.

To return to our neighborhood with a polluted river, we cannot achieve much of anything until we convince our neighbors that the polluted river is a problem. Perhaps some neighbors see the pollution as a positive development; the lack of wildlife may be appealing, particularly for those who are unaffected by health risks to children or pets. Those who live farther from the river may instead be focused on a different problem, like a pesky neighbor who throws late night parties or a road with too many potholes. The community at large can certainly address more than one issue at a time, but it cannot address *everything* at once. Choices need to be made and priorities need to be set. In order to be successful, we need to craft a conversation about the issue that makes clear the importance of the issue and the range of potential solutions. Until our problem and proposed solution are a part of the public agenda, the chances that we will overcome collective action problems to achieve some remedy are extremely low.

Collective action, institutional design, and agenda-setting are interconnected concepts and together they determine much of how we, as individuals within social groups, can explain our lives as well as the political actions that affect us. Overcoming collective action problems may depend upon our ability to shape the public agenda and convince others that the change we seek is relevant, important, and viable, not just for us, but for them as well. Even if we succeed in shaping the agenda, however, a lack of institutional knowledge or an unfavorable set of procedures may stymie our efforts. And even a favorable institutional environment does not guarantee success if we cannot hold together a group of people who agree on the goals we seek. When we complicate the matter further by considering different levels of government (local, state, territorial, national, international) and the multiple institutional settings (bicameral legislatures, separation of powers) that share responsibility for policy-making, the value of understanding these concepts becomes abundantly clear.

What Makes Political Science Interesting and Important?

What makes political science interesting and important can be stated as a simple fact: it is virtually impossible to conceive of a situation that does not implicate these basic theoretical understandings of social life.

My own career in the field of political science has been rather unconventional. As a young girl, I was fascinated by the politicians that I saw on TV – they seemed knowledgeable, powerful, and important. Working on campaigns during college, I quickly learned that the electoral system is, in practice, messy and imperfect. Internships with the Maryland Governor and a think tank in Washington, D.C. allowed me to see more of the policy-making process and prompted me to consider how issues are defined, how elected officials work with the public, and how much good is achieved by those who never appear on TV but work tirelessly behind the scenes. During graduate school, I had a chance to work at a prominent Washington think tank, completing research and interacting with a diverse range of political voices.

Today, I am a tenured professor at Drake University, where my primary love is teaching. Over the past decade, my teaching responsibilities have been complemented by ongoing involvement with and appreciation for organizations that seek to engage the political world through research and advocacy, with particular focus on practical strategies to increase public participation in government. In these capacities, I have the opportunity to travel throughout the country and internationally, meet exceptionally talented, innovative and inspiring thinkers, examine policy successes and failures, think creatively and work to solve problems.

More than I ever realized during my initial years of study, the field of political science has defined the way I think about the world. It's a discipline informed by critical thinking, broad systemic inquiry, and appreciation of diverse views. The tools and modes of thought associated with political science can find application in any and all areas of our social life including business, law, media, fine art, science, advocacy and social change, and yes, government.

Politics is Everywhere Around us

Too often, "politics" is misrepresented as either (a) a simplistic game whereby brute power is exercised irresponsibly; or (b) an abstract and untouchable world of elites in distant locations. And too often, "political science" is conceptualized as either (a) a training ground for future politicians, more ambitious than bright; or (b) an exclusively academic world where complex social dynamics are over-simplified and studied in a vacuum outside of public view.

Instead, I would propose that politics is everywhere around us, deeply embedded in even the most ordinary circumstances of our day-to-day lives. I would likewise propose that political science is an effort to systemically understand the dynamics of social interaction and the processes that produce collective outcomes. The sheer number and range of applications is itself exciting. In my own life, I have also found that the principles of political science, applied in a practical and intentional way, have empowered me to be a more thoughtful, informed, active, and effective citizen.

Chapter 3: Showing How the Magician Does A Trick - Chapman Rackaway, PhD

Chapman Rackaway, PhD
Professor of Political Science
Fort Hays State University

Dr Chapman Rackaway serves Fort Hays State University as Professor in the Department of Political Science. Professor Rackaway is also the advisor for Master's students in Political Science, Public Administration, and Political Leadership and Public Service. In 2007, 2010, and 2013 Rackaway was a nominee for the Pilot Award, FHSU's highest award for teaching faculty. Professor Rackaway is an avid believer in the role technology plays in both teaching and politics. Rackaway uses instant polls, web video, lecture capture, interactive graphics, and freeware tools to engage students in course materials. A recognized expert in Kansas politics, Rackaway is regularly quoted in stories about politics Wichita Eagle to the New York Times. A frequent guest on KAKE-TV's This Week in Kansas, Rackaway has established a statewide media presence as a member of the Insight Kansas consortium of political scientists who write weekly columns about Kansas politics, Rackaway is syndicated in more than thirty newspapers throughout the state. Rackaway also hosts Talking Democracy, a weekly public-affairs television show on KFHS television and is a regular moderator for Smoky Hills Public Television's Kansas Legislature program and Kansas Candidates debates.

"Why did They do That?"

Political science is a unique academic discipline because it provides the student an opportunity to pierce the veil of one of the most opaque elements of human behavior: politics. Few disciplines focus their attention on a subject matter that is as omnipresent in the modern citizen's life or impactful as human governance.

For those curious about the vagaries of human interaction, there is no better field of study. Political science allows us to see beyond the theatrics and rhetoric of a political campaign and examine what the campaigns are really trying to accomplish. For students with an eye to working in politics, studying political science will let them develop a deeper understanding of human behavior and give them the tactics to convince and persuade those voters when on the campaign trail.

Think about any time you have watched a political ad or consumed some political news. You likely wondered, "why did they do that?" Why did Republicans support shutting the government down on multiple occasions, or why did Vice President Joe Biden famously exaggerate in the 2012 presidential election campaign by stating to an assembly of African-Americans that his ticket's opponents wanted to "put [them] back in chains?" Political figures often say things that are inflammatory, extreme, or just downright silly. Candidates and elected officials say and do such extreme or odd things for specific reasons, ones that are not reported in the news but students of politics can elicit. Politics has a language and culture of its own that hides the real tactics of a campaign message from journalists and the public. (McCutcheon and Mark 2014)

Political Science is Like Watching the Movie "The Matrix"

Another way of thinking about political science is like watching the movie "The Matrix." In the movie, the character Morpheus invites Neo to select one of two pills: the blue pill will return Neo to his previous life, comfortably oblivious to the real machinations of the world behind what he saw on a daily basis. The other pill, the red one, would awaken him to the real life that exists beneath the illusory surface that he had previously accepted as reality. Studying political science is taking the red pill: you can discover the rationales for political rhetoric, understand the decisions made by elected officials, and see through the veneer of campaign messages through the study of politics and elections.

At Fort Hays State University, we have a specific program within the political science department that trains students to become political professionals: the Political Management concentration within our political science major. In addition to traditional classes in Political Behavior and Political Parties, we offer classes in Lobbying, Political Communication, and Political Campaign Management. As a former political professional myself, I teach students the process of planning campaigns, designing messages, and executing political operations from a place of theoretical knowledge and experience. Students who want to run for office one day or wish to be a leader on a political campaign can get practical knowledge in addition to the theoretical.

Having worked in politics from designing websites and polls to managing entire campaigns, I have seen how strategy works and why campaigns often do things that seem illogical or ill-advised. Another and better way of looking at it is similar to how the performer Penn Jillette presents his shows. Jillette began as a magician, and magicians have a professional code mandating they do not share how they perform their tricks. Jillette, though, does exactly that: he shows how he does the tricks and how other magicians use simple sleight of hand and misdirection to fool the audience. I compare my teaching to Jilette's approach, because I teach students how the magicians do their tricks.

Much of Politics is Theater

Much of politics is theater. By that, I mean that the content we consume is scripted and specifically designed to achieve goals. A political figure rarely speaks straightforwardly and honestly, instead hedging and obfuscating what they actually mean. Citizens unfortunately have demanded this. We know that citizens pay precious little attention to politics, generally start paying attention late in a campaign, and are susceptible to messages that play on fear or other anxiety. The rhetoric campaigns present is not intended for the involved and well-informed. That language is focused on engaging and motivating the disconnected voters who will only start paying attention in the last two weeks of a campaign.

A helpful comparison here is in the area of illegal immigration. America's borders are not secure and millions of undocumented foreigners are living and working in the United States. What should be done about this important policy issue? During the 2012 elections President Obama and Republicans like Mitt Romney traded extreme ideas like completely open borders or 'self-deportation'. Why do politicos not moderate their language and meet in the middle? Because the things they want to say would likely alienate most of the soft potential late-deciding supporters. The voters who begin to pay attention late are the most susceptible to campaign messages. (Fournier et al 2003)

The theater of politics is never greater than in the campaign environment. Early political campaigns in this country were straightforward mobilization efforts by political parties. At a time when ninety percent of the public aligned with one of the two major political parties, campaigning was relatively easy. Mobilizing the base was vital, so organized rallies and ballot stuffing comprised the primary foci of campaigners.

But the Progressive Reforms of the early 1900's changed everything. Political parties lost their ability to nominate candidates directly, fundraise as broadly, promote their issues and candidates through party-owned newspapers, and generally organize elections. When the parties lost their power, voters lost their need to connect to politics through parties, and so party identification dropped from ninety-percent to just over fifty percent and the campaigners had to change their tactics. Mobilization was important still, but only part of the process. Campaigns had to persuade, and the persuasion element in today's campaigns is where most of the theater can be found.

Why do Campaigns Consistently 'Answer' Questions...?

Today's professional and persuasive campaigns are different. Message creation and discipline are of paramount importance. Why do campaigns consistently 'answer' questions or accusations by diverting attention back to the theme of their campaign? Because campaigns know they get an incredibly brief window of time to communicate their message to the voters. And the message, the core piece of information the campaign wants people to know if they know nothing else, has to be consistently presented to become ingrained in the voter's mind.

Campaigns carefully craft messages to minimize negative perceptions of their candidate, maximize his or her positive, and compare their candidate as a superior option to their competition. Doing so in six seconds – the average length of a televised news soundbite on a campaign - is incredibly difficult and leads campaigns to make hard choices about what information they share and what they choose not to say.

Take a staple of the modern campaign: the 'negative' ad. I put negative in quotes because the term is a loaded pejorative that places focus on the fact that the ad contains a message that puts the sponsoring candidate or group's opponent in a negative light. The general public complains constantly about such ads, so you would expect that the public backlash against them would lead to campaigns not using them. But 'negative' or comparative ads persist. Why, if they are so unpopular?

Political science research gives us the answer: because while the public may talk a good game about the ads, comparative ads are the ones that voters pay attention to and those ads result in the greatest level of learning, recall, and change in candidate support levels. In other words, comparative ads are so prevalent because they work. (Ansolabehere and Iyengar 2010)

Another helpful example of a difference between what we see available for public consumption and reality is in campaign messages. In my adopted home state of Kansas, the 2014 gubernatorial election was a hotly contested one between incumbent Republican Sam Brownback and his Democratic challenger, Paul Davis. Brownback's primary campaign message about Davis was that he was a supporter of President Barack Obama. Why would the Brownback campaign try to make the election about Obama instead of his own record or Davis'? Because Brownback, who presided over steep tax cuts and declines in state revenue during his term in office, was unpopular and trailing in most polls. Davis as a mostly unknown quantity had little record to attack. But the Brownback campaign had conducted polls, finding that the most unpopular elected official in Kansas was none other than President Obama. So the Brownback campaign tried to tie Davis to Obama as a method of undermining moderate and Republican support for the Democrat. Ads showing Davis with Obama and in support of him might seem to be irrelevant, but once one understands campaign strategy and tactics the approach makes more sense. Davis' self-described record might be abstract and distrusted by voters, but the Obama comparison makes for an easy analogy to process mentally, and personally relevant and simple-to-decode messages are the most effective ones. (Popkin 1993)

Social Media is more than Just Outreach

Social media is yet another area of the campaign environment where the reality is not what is appears to be on the surface. Campaigns from President to county commissioner have Twitter feeds, Facebook profiles, and YouTube channels. Social media is more than just outreach, though. While campaigns might post silly tweets or inflammatory statements on their Facebook walls, they are doing more than just pushing content to consumers. Beginning with the 2008 Obama Presidential campaign, campaigns began aggressively using social media, but not all of the campaigns used it to maximum effect as the Obama effort did.

As Sasha Issenberg detailed in *The Victory Lab*, the Obama campaign used social media content as a method of collecting massive amounts of data on users. The data collected by the Obama campaign informed them of what type of content drove the highest levels of engagement, was shared most often, and what was most directly related to a tendency to be amenable to direct contact by the campaign and voting. The real reason for a campaign social media presence is less about the content, then, and more about the data it provides and the deeper level of knowledge of voters that can be gleaned from that data. (Issenberg 2012)

The last example of seeing behind the theater of politics using political science is the way messaging is rhetorically constructed. Why do campaigns play on fear in their messages rather than rely on more technical, clinical issues-based appeals? Much for the same reason negative campaign messages prevail: fear works. Campaigns play upon those feelings of fear. Republicans worked fear over Obamacare in 2010 as a central strategy in their campaign to win partisan control of the House of Representatives during that election cycle. Fear, anxiety, and anger are mobilizing emotions, and understanding why campaigns use those tactics helps us understand them and realize that public statements might not reflect reality. (Brader 2006)

Campaigns do things for specific reasons, but not necessarily the reasons shared for public consumption. Political campaigns often engage in behaviors that make the public think those campaigns are incompetent or insane, but in fact the messages that come out of most campaigns are very carefully calculated. Studying political science, showing the structural constraints and incentives in campaign behavior, helps us strip away the theatricality and see the true motives behind the public statements. In other words, knowing how the magician does a trick lets you see the real truth and not the illusion. Since politics is so central to our lives, stripping away the deception is important for the citizen and a special opportunity for the scholar. For the curious mind, the ability to decode –or craft - political messages is a unique element of the study of political science.

References

Ansolabehere, Steven and Shanto Iyengar. 2010. *Going Negative*. New York: Simon & Schuster.

Brader, Ted. 2006. Campaigning for Hearts and Minds: How Emotional Appeals in Political Ads Work. Chicago, IL: University of Chicago Press.

Fournier, Patrick; Richard Nadeau; Andre´ Blais, Elisabeth Gidengil, and Neil Nevitte. 2003. "Time-of-voting decision and susceptibility to campaign effects." Electoral Studies. 23:661-681.

Issenberg, Sasha. 2012. The Victory Lab. New York: Basic Books

McCutcheon, Chuck and Davis Mark. 2014. Dog Whistles, Walk-Backs, and Washington Handshakes: Decoding the Jargon, Slang, and Bluster of American Political Speech. Washington, DC: ForeEdge.

Popkin, Samuel. 1993. *The Reasoning Voter*. New York: Knopf.

Chapter 4: Study and Research in the Discipline of Political Science – Professor Michael Nojeim, PhD

Michael J. Nojeim, Ph.D.
Associate Professor and Program Coordinator, Political Science
Prairie View A&M University

Dr. Michael J. Nojeim is Program Coordinator and Associate Professor of Political Science at Prairie View A & M University. He received his Ph.D. in International Relations from American University in Washington, D.C. He teaches a wide range of Political Science courses including American Government, International Relations, Comparative Politics, U.S. Foreign Policy, Gandhi and King, and Political Studies through Film. His areas of specialization include Nonviolence, Middle East Politics, and U.S. Foreign Policy. He has published scholarly articles on international relations and on nonviolence. His first book, ***Gandhi and King: The Power of Nonviolent Resistance, was published*** *in 2004. His second book, co-authored with David P. Kilroy, is called* ***Days of Decision: Turning Points in U.S. Foreign****, and was published in 2011. His third book, co-authored with Frank D. Jackson,* ***Down That Road: A Pictorial History of Prairie View A&M University****, was published in 2011. Dr. Nojeim also actively serves the academy, the university community and the Houston area as well. He was a member of the editorial board of the International Journal of Gandhi Studies. He has served as the Chair or member of many Division and University committees, such as the President's Annual Symposium Committee, the University Core Curriculum Committee, the Division Strategic Plan Committee, the Division Accreditation Committee, multiple Faculty Search Committees, and many more. He also served as President of the Board of the Decade of Nonviolence—Houston from 2005 – 2009. He is also the Faculty Advisor to several student clubs and organizations on campus, such as the the Model United Nations Club.*

What is the Academic Discipline of Political Science?

In the simplest and most general of terms, an academic discipline is a field of study devoted to a particular subject, such as mathematics, biology and, of course, political science. Political science is concerned with understanding the structures and processes of government units, like parliaments or presidents, and those non-governmental institutions that connect people to their government. We call those non-governmental units linkage institutions and they include the media, public opinion, interest groups, political parties, voting, elections and campaigns. Political scientists uncover emerging trends in the media's coverage of politics and government and also study patterns in voting and elections.

For instance, they have found that voting patterns are heavily influenced by changes to the economy but less so by changes in crime rates.[1] Graduates in political science often find employment in linkage institutions, like interest groups and political parties, where they work on public policy issues whose impacts are felt across many economic and social sectors.

Studying political science is critical in the modern and global age because nearly all human activity is influenced by politics and government. Our economic well-being, our personal safety, our national security, and of course, the quality of our all-important educational systems, are all connected to proper government: and without good political scientists to assist in ensuring good government, nearly all of our human endeavors are bound to suffer greatly.

Three Critically Important Skills in Political Science

Faculty- as well as student-scholars who focus on a subject or discipline are expected to possess specialized knowledge in that area. Such specialized knowledge should provide graduates with at least three critically important skills.

---

[1] Michael R. Hagerty, "Quality of Life from the Voting Booth: The Effect of Crime Rates and Income on Recent
U.S. Presidential Elections," *Social Indicators Research* 77 (2006): 197 – 210.

First, they must be able to *explain* why something in their discipline occurred. For instance, political scientists are often called upon to explain why a war happened, or why the crime rate is going up or down. They might be called upon to explain the shifting opinions and attitudes of the public on important topics, such as capital punishment, abortion rights, or same sex marriage. Or they might be called upon to explain why there has been a shift in the balance of power between, say, the United States and China. The ability of political scientists to explain why something occurred will help policy makers avoid repeating previous mistakes.

Second, experts in a discipline should be able to *predict* what might happen in the future so that government and society can be prepared for what comes next. Political scientists are often called upon to predict when and against whom the next war will be, when and where the next terrorist attack will occur, who will win the election, and how the Supreme Court will rule on a controversial subject.

Or they might predict what might happen in Asia as China's power continues to increase. Such predictions form the basis of planning and preparation. However, if there is poor planning, then there is poor preparation for what might happen next: not only do politicians pay a grievous price for this, but so too will the citizens. Consider the fallout from the poorly planned American conquest and then occupation of Iraq in 2003 or the incredibly inept governmental response to Hurricane Katrina in 2005 that devastated New Orleans and much of the U.S. side of the Gulf of Mexico. The latter was the result of a colossal failure in planning at all three of levels of government in the United States: federal, state and local. We must therefore instill in our students the notion that "prior planning prevents poor performance" and emphasize how vital these "Five Ps" are for political scientists to help government plan for what happens next.

Third, experts in political science must be able to *prescribe* solutions for key domestic and international problems. In this political scientists are asked to recommend the best course of action to take to resolve an important issue of public concern. Political scientists are often called upon to prescribe solutions for preventing war (should the solution be military alliances, increased trade relations or joining international organizations?), for improving economic performance (tax cuts or increasing government expenditures?), for improving the quality of the people's health (Obamacare?), or for reducing the crime rate (tougher prison sentences, more rehabilitation programs or more police officers on patrol?). A major area of concern today is prescribing how other

Pacific powers, such as Australia, India, Japan, and the United States should react to China's rise?[2]

In addition, political scientists offer prescriptions about how government itself should be structured. This is vital because as was mentioned above, government decisions and actions have huge impacts on virtually every aspect of peoples' lives. Political scientists must answer key questions in order to prescribe solutions. What is the best form of government? Should we recommend a Greek-style citizen democracy? Should we recommend a constitutional monarchy? Should we adopt a presidential or parliamentary system? What type of economic model should be prescribed? Socialist? Capitalist? A blend thereof? And how should government institutions themselves be structured? Should there be a federal system (like in the United States and Australia) where the central government and the sub-national governmental units both share and compete for power and jurisdictional authority? Or should we prescribe a unitary government, like in the United Kingdom, where the central government in London makes most decisions and dictates policies to lower-level governing units?[3]

These are not easy questions to answer. And since there is no universally correct choice, political scientists must be able to prescribe different forms of government for different peoples in their diverse political contexts. For instance, it might be wise to recommend federalism for a country that is geographically large and possessed of regions with significant religious, ethnic, cultural and linguistic differences. A federalist structure that provides a degree of autonomy or self-governing authority to such disparate regions might be the perfect recipe for domestic tranquility as it helps avoid violent, separatist eruptions. Political scientists, therefore, are charged with keeping the peace, long before any mention is made of sending in the "UN Peacekeepers!"

What can a Person who Earns a Degree in Political Science Actually Do?

Explaining, predicting and prescribing in political science are tasks that should be based on empirical and theoretical arguments. Empirical arguments are

---

2 For an interesting article that seeks to *explain* China's rise, *predict* the reactions and then *prescribe* solutions, see Namrata Goswami, "Power Shifts in East Asia: Balance of Power vs. Liberal Institutionalism," *Perceptions* 18, no. 1 (Spring 2013): 3-31.

3 For different views on federalism, see Gordon DiGiacomo and Maryantonett Flumian (eds.), *The Case for Centralized Federalism* (Ottawa: University of Ottawa Press, 2010) and also Ruth Hubbard and Gilles Paquet (eds.), *The Case for Decentralized Federalism* (Ottawa: University of Ottawa Press, 2010).

based on the collection and analysis of data and other forms of evidence. Theoretical arguments are based on assumptions, axioms and logical argumentation and conclusions, but they too should be built upon verifiable evidence. At large American universities, undergraduate students who major in political science are sometimes called upon to concentrate in one or two sub-fields within the discipline. A simple google search of diverse U.S. universities reveals several common sub-fields, including: American politics, political theory, political methodology, law, public policy, public administration, comparative politics and international politics. In each of these sub-fields, both empirical and theoretical training occurs.

Studying political science provides graduates with broad-based tools such as research, writing, and critical thinking skills. A degree in political science arms graduates with problem-solving skills that make them competitive in the job market. Many people ask "what can a person who earns a degree in political science actually do?" Well, what *can't* they do? Graduates in political science possess skills that are applicable to virtually every professional endeavor.

Indeed, many famous U.S. politicians, such as Hilary Clinton, Condoleezza Rice and Barack Obama, earned degrees in Political Science. But so too did actor Alec Baldwin, sportscaster Greg Anthony, AOL CEO Steve Case, and Nobel Laureates Ralph Bunche and Elinor Ostrom. Really, the possibilities are endless.[4]

Moreover, an advanced degree in the field—a master's or doctorate—provides specialized training for several particular professions; acquiring an advanced political science degree helps make the graduate even more competitive on the job market. An advanced degree in public policy, for example, might lead to employment as the Legislative Aide to a Member of Congress or it might lead to a job as a Policy Analyst for the Environmental Protection Agency. It might also lead to employment in one of the more than 40,000 public interest groups found in Washington, D.C. alone. Often called lobby agencies or non-profits, these interest groups range from civil rights organizations, like the National Association for the Advancement of Colored People (NAACP), to environmental groups like the Sierra Club, to animal rights advocates like People for the Ethical Treatment of Animals (PETA), to public safety organizations, like Students Against Drunk Driving (SADD).

---

[4] American Political Science Association, "Careers in Political Science," (n.d.), http://www.apsanet.org/content_6457.cfm.

Similarly, students with advanced degrees in public administration are trained in government management, economics and budgeting. They can find work as a City Manager, which is a professional—not political—position hired by elected officials to run the day to day operations of the city. For those who earn a Master's Degree in International Relations, there are government jobs in the Foreign Service, the Peace Corps, the International Trade Commission and so on. There are also private sector positions in journalism or as a staff member for international non-government organizations (NGOs), such as Greenpeace or Amnesty International. Working for an NGO involves publishing academic research that will influence public debate on important social and political issues. Amnesty International, for example, publishes studies showing how ineffective capital punishment is on lowering violent crime rates.[5]

Historically, the "jewel in the crown" of advanced degrees for political science undergraduates is the J.D., the Juris doctorate. For many of my students, this is their Holy Grail. But be advised: becoming an attorney is not so easy. The selection process for admittance to top law schools is quite competitive. Moreover, the employment prospects for recent law school graduates in the United States are not as promising as they were for law school graduates in the 1990s.[6]

The Terminal Graduate Degree in Political Science

The terminal graduate degree in political science, the doctorate of philosophy (Ph.D.), takes years to complete and is a rare achievement. In the period spanning 2002 – 2012, only 10,414 people in the United States earned a doctorate in political science, or in a related field like international relations.[7] Those who have earned their doctorate find employment in the types of fields

---

[5] Amnesty International, *Not Making Us Safer: Crime, Public Safety, and the Death Penalty* (London: Amnesty International Publications, October 2013), https://www.amnesty.org/en/library/asset/ACT51/002/2013/en/2efaa0f6-6899-459d-93e1-4bfe20004017/act510022013en.html.

[6] See for instance Adam Cohen, "Just How Bad off Are Law School Graduates?" *Time* (March 11, 2013), http://ideas.time.com/2013/03/11/just-how-bad-off-are-law-school-graduates/ and "The Price of Success: Some American Law Schools are Paying many of their Graduates' Salaries," *The Economist* (March 15, 2014), http://www.economist.com/news/business/21599037-some-american-law-schools-are-paying-many-their-graduates-salaries-price-success. Cohen cites data indicating that less than 2/3 of 2009 law school graduates actually got jobs for which they needed to pass the Bar Exam and *The Economist* article says that demand for law school graduates is down by 30% in recent years.

[7] National Science Foundation, "Doctorate Recipients from U.S. Universities: 2012" (Arlington, VA: National Center for Science and Engineering Statistics, December 2013), http://www.nsf.gov/statistics/sed/2012/start.cfm.

and organizations described above, but usually at a higher rank and salary. However, people who earn the doctorate in political science are usually interested in an academic career as professors at a university where they are expected to excel in teaching, research and service to the university and community. In fact, for most professors, publishing research is critical to obtaining tenure, or job security.

But political science research is most valuable when it helps politicians, government officials and industry officials do their job better. Political science research that has practical applications and solves real-world problems—such as inflation, unemployment, or terrorism—is absolutely vital to human progress. Do tax breaks for the wealthy help or hurt the economy? What caused the economic meltdown in 2008 – 2009 and what steps should we take to avoid a similar calamity in the future? Do international trade agreements such as NAFTA help or hurt a domestic economy's employment and inflation indicators? What government policies increase the domestic economy's competitiveness in the global economy? Do arms buildups prevent or cause military conflicts?

These are absolutely vital questions for which policy makers and other government officials must have answers in order to make the correct decision. The properly trained scholar in political science can help answer these questions by collecting data and analyzing evidence and then making recommendations.

Academic Research in Political Science

Academic research in political science, if grounded on reliable and verifiable evidence, can be incredibly useful to policy makers. For instance, consider the death penalty, which a declining but still significant majority of Americans favor.[8] Those who support capital punishment often argue that the death penalty deters crime. But is such a theoretical argument actually true? Government does many things: it builds schools, paves roads, and provides assistance to the needy. But is there anything more serious or more profound than government taking the lives of its own citizens in the name of law and order? How could such a practice be implemented without knowing whether or not it actually works? To be sure, the death penalty is one government policy

---

[8] Jeffrey M. Jones, "U.S. Death Penalty Support Lowest in More Than 40 Years: Sixty Percent of Americans Favor Death Penalty for Convicted Murderers," Gallup Poll (October 29, 2013), http://www.gallup.com/poll/165626/death-penalty-support-lowest-years.aspx.

that cannot be undone. Government cannot afford to make a single mistake on such a weighty matter for if it does, it would threaten the foundation on which government legitimacy is based and in whose name the people place their trust and loyalty.

Indeed, government is really the only institution that is considered to possess a monopoly on the legitimate use of force, including taking the lives of its own, or another country's, citizens. As such, government had better "get it right" when it comes to killing. And this is where political science researchers MUST play a critical role. For they are charged with testing the theory of whether or not government killing—in domestic or foreign affairs—is sound public policy.[9] Does the severity of the punishment—say, for instance execution instead of life imprisonment—work to deter the hardened criminal from committing a murder? Or does the sureness and swiftness of punishment, regardless of severity, have a greater deterrent impact? It is through their research on these matters that political scientists can help *explain* what *causes* lower homicide rates and thereby assist politicians or other government officials in making informed decisions about the correct course of action.

To be sure, all levels of government in the United States hire political scientists to conduct research in order to assist in making important decisions. Earlier in my own career, after completing a Master's Degree in International Affairs, I worked for a private contracting firm on behalf of policy makers in the American defense and intelligence agencies. We specialized in studies connected to terrorism. We researched terrorist grievances and motives, targeting strategies, recruitment strategies, and propaganda strategies. All such studies were so that government policy makers—in both civilian and military departments—would be correctly informed on the best course of action regarding anti-terrorist policies. Such research has real-world implications for the safety of countless innocent civilians.

After completing a Ph.D. in International Relations and working as an academician, my own research interests have been varied, but much of my published work is concerned with nonviolent alternatives to violent forms of conflict resolution.[10] I examine ways of managing, and resolving, conflicts

---

[9] For a collection of essays that examine this issue from different points of view, see Robert M. Baird and Stuart Rosenbaum, eds., *Death Penalty: Debating the Moral, Legal and Political Issues* (Prometheus Books: 2010).

[10] See Michael J. Nojeim, Gandhi and King: The Power of Nonviolent Resistance" (Westport, CT: Preager, 2004) and also Michael J. Nojeim, "A Gandhian Blueprint for Nonviolent Social Change," *Peace and Change* 32, no. 4 (October 2007): 546-573.

using means other than violence. So long as people have values that clash, there will be conflict. So, I do not try to find ways to "end conflict once and for all," for I find that an unrealistic and overly idealistic endeavor that serves neither the academy in its lofty ivory tower nor the tangible world. Discovering nonviolent alternatives that will help resolve a conflict as peacefully as possible is a relevant and worthy goal for a political science researcher. My research presents nonviolence as a rational strategic or philosophical alternative to riots, terrorism and war. Building on the examples of Mohandas K. Gandhi in India and Martin Luther King, Jr. in the United States, my work seeks to understand the ways Gandhian and Kingian resistance might be applied or replicated in other contexts. For instance, *Otpor!*, a nonviolent student youth movement in Serbia, helped bring down Serbian dictator Slobodan Milosevic at the turn of the century.[11]

Best Practice Issues for Professors and Students

In addressing the "best practices" of research in the discipline of political science, research is best when it reinforces academic teaching and teaching is best when it reinforces research. Research and publication for its own sake is rarely a productive enterprise as such leads to rather esoteric publications that are read by few and have an impact on fewer still. Rather, research that informs the educational enterprise, and is integrated into the classroom whereby it involves students, holds the most potential for it carries with it a dual advantage: on the one hand, it can produce original, relevant knowledge, even as it sparks the students' intellectual curiosity on the other. A professor charged with teaching the nonviolent resistance techniques of Mohandas K. Gandhi and Martin Luther King, Jr. will improve his or her pedagogy by virtue of researching and publishing on the very same subject matter. And students will be the direct beneficiaries for they can be involved in education not only as passive receptors of established knowledge but also as active participants in the creation thereof.

Professors should not use or exploit students to advance their personal research agendas. Rather, the best practice seeks to integrate and involve students in the research experience as much as possible in which the student is fully invested in a partnership with the professor who acts as the "guide on the side" and not "the sage on stage."[12] Consider: does a student learn better when

---

[11] Bryan Farrell and Eric Stoner, "How We Brought Down a Dictator: 10 Years Later: A Conversation with Srdja Popovic, a Leader in the Nonviolent Overthrow of Slobodan Milosevic," *Yes!* (October 7, 2010), http://www.yesmagazine.org/peace-justice/how-we-brought-down-a-dictator.

she merely takes notes, memorizes dates and other information, and then regurgitates this back to the professor on an exam? Or instead, does she retain greater depth and breadth of knowledge if she is called upon to actively participate in a research experiment, create knowledge that answers relevant questions and then present that new knowledge to her peers, whether in front of the class or at an actual academic conference?

Perhaps one illustration shall suffice. The U.S. government sponsors a program called "Research Experience for Undergraduates," (REU) wherein qualified university students are admitted to an intense "research boot camp" that is highly competitive with less than a ten percent acceptance rate. Once admitted to the REU, students spend the summer conducting original research with the goal of presenting their findings at conferences and ultimately publishing their work. Undergraduates are paired with a faculty mentor in accordance with their shared research interests. Together they develop a hypothesis, construct a research design, collect data, test the hypothesis against that data and draw conclusions. Prairie View A&M University has had three students admitted to this program in the last four years, all of whom have produced and published original research that has not only greatly enriched their undergraduate experience, but has also made a real contribution to the stock of knowledge.[13]

For instance, one of our student's research projects demonstrated a connection between education levels and domestic unrest in Global South countries: alarmingly, she found that the more educated a youth population is, combined with few employment opportunities, the more likely there is to be domestic unrest in the form of urban riots, terrorism and even guerrilla war.[14]

The point is that research should be conducted with this dual purpose in mind: that which benefits the student's own intellectual growth and that which benefits the real-world. The professor's interests should be secondary to those goals. Ultimately, therefore research programs should be driven by teaching. And teaching should be driven by passion. In such fashion may a professor's passions inform his or her students and hopefully infect them with the same desire to learn from the existing knowledge and then the curiosity to create

---

[12] Allison King, "From Sage on the Stage to Guide on the Side," *College Teaching* 41, no. 1 (1993): 30 – 35.

[13] See for instance, Nia Yarborough, "The Relationship Between Foreign Direct Investment and Intrastate Conflict," *The Eagle Feather* (2012), http://eaglefeather.honors.unt.edu/2012/article/14#.U4sfrSiooTB.

[14] Amber Smith, "Education and Civil Unrest," *The Eagle Feather* (2011), https://eaglefeather.honors.unt.edu/2011/article/45#.U4sfXSiooTA.

new knowledge. That, and not publishing for its own sake, is the highest reward a political scientist can earn.

Chapter 5: Political Science: More than Practical Politics – Professor Dinorah Azpuru, PhD

Dinorah Azpuru, PhD
Associate Professor of Political Science
Wichita State University in Kansas

Dr Dinorah Azpuru is associate professor of political science at Wichita State University in Kansas, United States. Previously, she was visiting assistant professor of political science and research coordinator of the Latin American Public Opinion Project (LAPOP) at Vanderbilt University. She continues her involvement with the LAPOP project, in which she is a member of the Scientific Support Group. In Guatemala, she was adjunct professor of political science and chair of the department of political science at the Catholic university of Guatemala (Universidad Rafael Landívar) as well as senior researcher at ASIES. Azpuru has a PhD in Political Science from the University of Pittsburgh and other graduate studies in Guatemala, Sweden and Switzerland. Her expertise is on democratization and democracy assistance to developing nations, survey research in Latin America, U.S.-Latin America relations and peace processes in Central America. She has published extensively on those topics in journals such as Latin American Politics and Society, the Journal of Democracy and Orbis among others. She is co-author of ***Construyendo la democracia en sociedades posconflicto: Guatemala y El Salvador en perspectiva comparada*** *and is currently working on a book about authoritarian political culture in Latin America. She teaches courses on comparative politics and research methods in political science.*

How can Politics and Science be Together?

When people hear that someone is studying political science they tend to think that they like practical politics—either they want to become politicians or they are interested in making a career running political campaigns. Furthermore, since politics is discredited in many countries, people may wonder why would someone choose to get a major and moreover a graduate degree, in 'political science'. People may also be baffled by the word 'science'. How can politics and science be together?

While it is true that someone who studies political science may be interested in becoming a politician or in becoming a member of a politician's staff, the truth is that the study of politics opens multiple avenues for professional development.

Let's start with the fact that political science has four well-defined fields of interest: domestic politics (the political dynamics within the country where the university is located), comparative politics (studying how political systems and processes work in other countries around the world), international politics (studying how countries in the world interact with each other) and political theory (the study of the philosophers and thinkers who dealt with political issues). In some countries, public administration is another branch of political science, while in others it is a graduate-level degree; in fact, master programs in public administration are popular among students who have completed undergraduate degrees in political science.

Finally, in some countries, political methodology (which focuses on the methods that political science uses to arrive to conclusions) is considered a separate field in some universities, especially at the graduate level . However, it is important to note that increasingly universities around the world are requiring political science undergraduates to take a basic political research methods, since it applies to the different fields.

Students who major in political science will often have to take required courses in the four main fields mentioned above, and depending on the country, they may have the option of specializing in one of them by taking more courses in their field of interest. The fact that political science covers those four fields presents multiple opportunities for professional development, all of which go well beyond practical politics.

Even students who ultimately focus on the domestic politics of their own countries are not restricted to finding jobs as politicians or running political campaigns; they can alternatively become analysts for think-tanks or media outlets, work for non-governmental organizations or for different levels of government within their country. Those who focus on comparative politics or international relations have a wide variety of opportunities to get involved with inter-governmental or non-governmental international organizations and in a globalized world these two fields are of uttermost importance. The professional avenues for those who are interested in political theory is more limited, but teaching is one of the possibilities.

Regardless of which of the four fields political science students are interested in, they will graduate with a degree that allows them to have a broader perspective about their own country and the world than most other academic majors. In many countries, beyond courses in the four fields mentioned above, students are also required to take courses on economics, law and sociology, which furthermore expand their academic foundation.

All things considered, students who graduate with a degree in political science are equipped to make a major contribution to society, and their broad and analytical perspective can be extremely useful for governments, businesses and non-governmental or multilateral organizations.

Students of Political Science should not be Afraid of Numbers

When students think of political science, many believe that they chose a major in which they will not need to use numbers—and many are happy about not having to use them. However they must consider that political science is increasingly becoming a quantitative profession, one that deals with data and numbers. While many universities do not require students to take quantitative courses related to politics, it is desirable that most students do, as it will increase their understanding of the world of politics or international relations, as well as their prospects for employment.

Students of political science should not be afraid of numbers. They are not expected to deal with computer programming or to understand abstract formulas. Instead, quantitative political science means being able to interpret numbers and derive conclusions from statistical outputs, and at the graduate level, to learn how to test hypotheses based on statistical data or mathematical modeling.

Hypothesis-testing is what makes politics a science, comparable to other social or hard sciences. Political Scientists who engage in research in any of the four fields mentioned earlier, try to understand recurrent patterns of behavior or citizens, politicians, political parties, countries or even international organizations. They test through a variety of methods, the relationship between factors (or variables) and try to propose new avenues for research based on their findings. The research published in political science journals around the world makes an enormous contribution to the understanding of the world of politics in a scientific way.

One of the areas that is becoming increasingly important is the study of public opinion in a scientific way. Political public opinion studies go well beyond the commercial polls that measure electoral preferences. Public opinion studies within political science interpret, measure and analyze the immediate but also the long-term behavior, attitudes and beliefs of citizens. We must remember that at the end of the day, in any society, citizens are the most important component of politics. Particularly in democratic societies, politics is largely driven by citizens preferences.

How do we Study What Citizens Think and Believe?

How do we study what citizens think and believe? The answer is through public opinion surveys. And while it is difficult and expensive for undergraduate—or even graduate students or professionals in the field—to conduct their own surveys, especially at the national-level, public opinion data is nowadays widely available. Different universities, think-tanks and even commercial firms, make public opinion data available.

It is important to know that students can choose to use secondary or raw data to understand public opinion. Secondary data refers to the data that has been collected, processed and made available. The links to some of the main sources for secondary, comparative public opinion data are included in Table 1; some of the links lead directly to the interactive data. The table also includes specific data for certain regions of the world.

Table 1
Sources of Comparative Public Opinion Data

NAME	LINK
World Public Opinion	http://worldpublicopinion.org/index.php
Pew Research Global Attitudes Project	http://www.pewglobal.org/category/survey-reports/
Gallup World Poll	http://www.gallup.com/strategicconsulting/en-us/worldpoll.aspx
World Values Survey (interactive)	http://www.worldvaluessurvey.org/WVSOnline.jsp
Eurobarometer Standard (Europe)	http://ec.europa.eu/public_opinion/archives/eb_arch_en.htm
Americas Barometer (the Americas)[15]	http://lapop.ccp.ucr.ac.cr/Lapop_English.html
Latinbarometer (Latin America)	http://www.latinobarometro.org/latOnline.jsp
Afrobarometer (Africa)	http://www.afrobarometer.org/
Asian barometer (Asia)	http://www.asianbarometer.org/

Source: Prepared by author

Raw survey data is that which has also been collected by someone else (a firm or organization) and has been entered into a spreadsheet, usually using statistical software (such as SPSS, Stata or R), but unlike secondary data, it has not been processed. Users of raw data can "crunch" their own numbers and obtain the output they desire. Political science undergraduate students can learn the basic statistical terminology (as well as the statistical software required to analyze raw data) in different courses not necessarily related to politics, but it would be desirable and more effective to take a course in which the study of statistics id directly linked to the analysis of political issues. Some of the sources included in Table 1 also make raw data available upon request.

[15] The author of this article specializes in research on public opinion in Latin America and in addition to being Associate Professor of Political Science at Wichita State University, collaborates with the Americas Barometer at Vanderbilt University in the United States, as a Member of the Scientific Support Group.

In either case, whether using raw or secondary individual-level data, it is important to ensure that the organization that collected it is an objective one, and that proper sampling methods, data collection procedures and unbiased questionnaires were used.

Beyond public opinion data, students and professionals of political science can also use what is called aggregate data, which is data usually referred to countries or provinces within countries (compiled usually by international organizations) that can provide a window into a variety of issues directly or indirectly related to politics. One of those examples is the Transparency International Corruption Perception Index produced by Transparency International (http://www.transparency.org/research/cpi/overview), or the Governance Indicators produced by the World Bank (http://info.worldbank.org/governance/wgi/index.asp). There are multiple indicators that although not directly related to politics, can be used in political analyses, such as the data produced by the United Nations. This aggregate information can be of particular interest to students interested in comparative politics or international relations.

Two Trends you should Know about Political Science

There many other things that an article on political science could discuss, but this article focuses essentially on two aspects: (1) students should be aware that political science is a versatile field of studies that goes well beyond practical politics; and (2) political science is becoming an increasingly quantitative field and students should embrace the use of numbers, figures and statistics without fear, as there are multiple ways and sources which allows students to use numbers without being expert mathematicians.

The study of politics has always been and will always be relevant. No society in the world, authoritarian or democratic, can sustain itself without some degree of political interaction of the different actors within its borders, and in a globalized world, without interacting with international actors. Understanding how those national and international political processes work is essential and that is the goal of those of us who study political science. Those of us who engage in the study of politics appreciate the multiple possibilities for professional growth and truly enjoy trying to understand the 'puzzles' that politics represents.

Many of the ideas I have shared above are based on my own experience as a Political Science professor. While many universities have undergraduate and graduate Political Science degrees, at Wichita State University in Kansas (United States) we only focus on the undergraduate level. However we believe in the importance of teaching students the skills to analyze and understand political processes both from the qualitative and the quantitative perspective, and for that reason have made Political Research Methods a required course. We also have required courses in the four fields that were mentioned in this article, but students have the possibility of choosing most of their courses in their particular field of interest.

As most undergraduate degrees in the United States, it takes students four years to obtain the undergraduate degree. There are scholarships at the university level and at the level of the Political Science Department, particularly for students who have a good academic record. Many of our graduates pursue graduate studies in different fields at other universities in the United States or in other coutries.

Chapter 6: Confessions of an Aging Political Theorist: Apologia Pro Vita Sua – Professor Ernest J. Yanarella, PhD

Ernest J. Yanarella, Professor and Chair
Department of Political Science
University of Kentucky

Dr. Yanarella is a Professor and the Chair of the Political Science Department. His primary teaching and research interests include critical policy studies of the energy and environment, agricultural and ecological policy, and national security and arms control. He serves as the Associate Director of the Center for Sustainable Cities and Director of Environmental Studies Program in the College of Arts and Sciences. Dr. Yanarella's teaching and research interests align well with the mission of this program, and his experience as an administrator provide valuable guidance to the program.

Academic Discipline of Political Science

The study of politics has a long and treasured heritage that goes back to the early Greeks, especially Plato and Aristotle, and finds their precursors, Theseus and Solon, and their badly misunderstood competitors, the Sophists, participating in the laying political foundations of our collective existence. Drawing upon the dialectical method of Socrates, Plato employed the dialogue approach to excavate the meaning of the good society and justice, governance and the laws, the quality of statesmanship and true leadership, and the nature of political knowledge itself. To Aristotle, we owe a normative ideal of the polis or political society and a conception of human beings as political animals derived from his method of exposition and his formulation of the four causes.

To Samuel Taylor Colderidge goes the adage that everyone is born either a Platonist or an Aristotelian. While not strictly universal truth, it is probably a truism.

From these origins, the Western political thought has served as a great tradition that has experienced both continuity and innovation over its twenty-five centuries, being revitalized by the glory and expanse that was Rome and its legal and political foundations, the Hebrew and Christian streams of religio-political thought and practice, the early modern animating impulses generated by the Protestant Reformation and the Renaissance, the Janus-faced writings of Niccoló Machiavelli, the rise and impact of the classical liberalism and the political theory of possessive individualism uniting Thomas Hobbes and John Locke, the challenges of Jean-Jacques Rousseau's musings on the general will and the noble savage, the stunning utilitarian defense of liberty and liberal institutions of John Stuart Mill, the conflicting critiques of liberalism by conservative founder Edmund Burke and communist founder Karl Marx, and the baleful pessimism of Max Weber, whose political theories were dressed in sociological garb.

The entering student of politics must never forget that contemporary political science and the inquiry into political ideas and practices remain largely grounded in and defined by the first and enduring words of political theoretical titans who loved theorizing about political things and who sometimes involved themselves in the rough-and-tumble world of politicking. In so doing, aspiring political scientists will come to recognize that politics is both a constructed medium with normative anchors that can never be overcome. Nor can its moorings in the ideas, concepts, and theories of the above political theorists (and others not mentioned) be simply jettisoned. If we are not all little Platonists or little Aristotelians, latter-day political scientists still bear the traces and hidden ties to the legions of political thinkers who have preceded them.

For all of the research literature in comparative politics, today's empirical researchers have not developed this sub-field much further than Aristotle in some respects. For all of the empirical studies of power and its application, the perspectives on power opened up by Machiavelli and Weber have not been totally surpassed—indeed, in important ways, they remain an essential starting point to grasping the subtleties of power and its many masked expressions.

As international relations theorists pondering the origins and consequences of globalization in our late twentieth and early twenty-first century world should realize, the colonizing of the early modern world by European imperialist powers in the fifteenth century and the vison of the extension of the bourgeois powers in the nineteenth century offered by Marx and Engels graphically, almost lyrically, drawn in the "Communist Manifesto" require that processes of top-down globalization were evidenced in prior eras and serve as sites of comparative analysis and insight in engaging subtle investigation of our latest travails.

The Future of Political Science

This realization suggests that the future of political science will oscillate between theory and practice, theorizing and political engagement in ways that cannot be completely separated or placed on separate ontological dimensions. It was Lenin who argued that nothing was so practical as a good theory, just as Camus reminded us decades latter that by our action or inaction we too shall enter the political fray.

So too do the burgeoning possibilities and scope of careers for political science majors point to the continuity and separation of theoretical (or academic) careers and practical vocations in usual politics and politicking (political consultants, campaign managers and aides, "spin doctors," etc.) or less conventional politics and political activity (NGO actors, INGO practitioners, social movement leaders and strategists, citizen activists, new party advocates, etc.).

The political and social world of today is a data-driven and data-facilitated one. The true makers of the future are the political science knowers who connect the past, the present, and the possible future through research that alternately or in combination draw upon grounded knowledge of the past, trends and tendencies in the present, and alternative possible futures that all rest upon structured methods that can be learned and honed. These tools will help cast greater light on how the inheritances from the past inform and shape the present, how the residues of past social, political, and cultural forces stymie the hoped-for future, and how the unrealized hopes and aspirations of the present remain resources in a future that is simultaneously never completely blocked and never entirely open.

At the University of Kentucky, its political science faculty largely use quantitative methods to examine a wide range of contemporary problems with explicitly or tacitly normative ends in mind. Three of our international relations faculty study: domestic conflict/instability, coups d'état, regime types and democratization, and international education, as well as the modeling the rivalry process, the influence of International Organizations on conflict duration, and the politics of military coalitions. As against the adage to "study war not more," these three rising scholars believe that conflict resolution and the prospects for peace demand the study of international security, international institutions, and conflict processes.

Finally, a fourth faculty member among this group engages overlapping empirical issues and normative questions about the politics of the World Bank's and International Monetary Fund's structural adjustment lending and the impact of global economic integration on democratic governance, human rights and the welfare state.

Research Issues in Politics: Inequality, Crime, Security, Justice, Law and Sex

Our three comparativists move beyond regional studies research programs even as several of them base some of their work on Latin America by studying election-related protests and their consequences for democracy in developing countries, the links between economic inequality, poverty, and crime and their effects on ordinary citizens' political attitudes and behaviors, and conditions under which women's numeric representation will influence the substantive representation of women. Despite the misogynistic and anti-democratic prejudices of Aristotle, the original founder the comparative method, would, I think, find the latent normative imperatives driving the work of these three women faculty residing in an unbroken, but cumulative, tradition beginning with him.

Our Americanists, both institutionalists and behaviorists, contribute to research that remains linked to key questions about executive leadership, legislative process and the corrupting role of money in democratic politics, the interaction of domestic and foreign policy behavior, and elections and voting behavior with a focus on racial/ethnic politics. The judiciary as an institution and a research landscape for the study of law, politics, and justice is not slighted among this university's political science faculty. One of our judicial scholars studies judicial behavior and decision making, judicial process, oral arguments, Supreme Court confirmation hearings, and the role of courts in American society, while another investigates the effects of minority stress on the well-being of LGBT individuals and same-sex couples, and legal status issues for same-sex couples. Still another conducts research on the politics of the military and national security and law and sex.

The two UK political theorists are heavily involved in the nexus of political theory and public policy. My own work is borne of the professional self-identity of a critical policy studies researcher and instructor whose work has spanned early and modern political theory, peace studies, environmental and agricultural policy, politics and literature and, for thirty years, sustainability and particularly urban sustainability in North America, Europe, and, most recently, China.

The other political theorist has championed the study of Americana political thought and Appalachian politics with a driving concern to advance citizen participation and a more engaged form of democracy through the use of futuristic techniques initiated by Harold Lasswell and now being advanced by a new crop of futurists who draw upon refined and sophisticated tools of citizen engagement in imagining alternative futures operating below and beyond the stodgy processes of usual politics.

Maturing and successful political science departments achieve collegiality and camaraderie by striving for strong doses of intellectual tolerance, methodological pluralism, and a wide-ranging undergraduate curriculum and graduate studies program. As my department works to rebuild its foundations and its numbers, these are the ingredients I as chair seek to employ in constructing a disciplinary house and educational curriculum worthy of the vocation of political science.

Benefits of Academic Research in Political Science for Practitioners

The range of research into politics founded in the department that I chair can be found in department after department in the United States and Canada and bears kinship with the studies of British universities and many of the institutions of higher education in much of Europe. The overtly empirical studies of contemporary political science are sometimes criticized for their narrow focus often dressed up with seemingly arcane political methods whose search leads to streams of research reports and publications that appear catalyzed more by technical rigor than substantive significance. The truly prominent political scientists however are characterized by the motivations and aspirations of their past brethren of the Western political tradition to engage large issues and conundrums of politics and to better understand and improve the collective lives of everyday citizens.

While some students of politics content themselves with impacting the contemporary world of politics by serving as part-time consultants churning out the results of attitudinal surveys or those of focus groups for political candidates, many political scientists seek to transfer their political knowledge from research for the purpose of enriching the present-day polis, advising citizens action groups or environmental organizations, defending labor unions under siege, working for the cause of regional or even global peace, drawing up blueprints for sustainable cities, advancing women's causes, reversing social and economic inequality, improving the quality of our civic life, and protecting constitutional and civil liberties for oppressed minorities.

Surveys of the political science community in the United States have shown that American political scientists are disproportionately liberals in the citizen roles in politics. The assault on liberalism by the secular and religious right has pushed the center of US mainstream politics increasingly to the right to a point where the term "liberal" has become a pejorative epithet. It has increasingly prompted liberals to substitute "progressive" or "advocate for working people" as alternative framings.

If political science is not to retreat to the margins or safe havens of narrow professionalism and political quietism, liberal political scientists will need to come to terms with the need to redefine themselves as knowers and actors. One place where my research has taken me is to China, where the opportunities for exploring the scope and limits of political liberalization, the nascent signs of civil society, the halting steps toward the rule of law, the rising protests of inequality of economic opportunities in the era of market socialism with Chinese characteristics inaugurated by Deng Xiaoping, the persisting imbalance between high economic growth and sustainable urban and resource practices, and the continuing tight limits of democratic expression and practice within the hegemonic control of the party-state since Tiananmen Square are visible. In these studies, my thoughts are never very far from alarming developments in the United States.

Special Need for Political Analysts

At a time when corporate power and wealth and income inequality are reaching new heights and a new epoch of multinational and transnational capitalism triumphant appears near, the special need for political analysts and other social scientists to explore the range of economic alternatives and the politics of alternative economic institutions and practices both within and beyond capitalism is sorely needed. Given the arguably close association among relative social equality, engaged democratic institutions and practices, greater worker participation, and more cooperative economic institutions, the time is pressing for political students as researchers and actors to build upon the literature on Emilia-Romagna, Italy, Mondragon, Spain, and elsewhere in the present and the past as means of expanding the political imagination of economic institutions. The goal of this exercise will be to provide new options and future possibilities against the stifling trends toward transnational corporate capitalism and galloping, anti-democratic tendencies toward plutocracy.

Finally, any American political scientist pondering the fate of the commons in the realm of politics must recognize that climate change poses no less an issue for public policy and of political education, given the overwhelming consensus of climate scientists broadly understood. A profession of politics and political science that does not care about the ecological auspices of human life is not a vocation that is not fully capable of rationally defending its humanistic concerns in the face of the increasing corruption of our political commons or the increasing deterioration of the ecological commons of our earthly host and partner that has for thousands of years maintained the conditions for life on this planet and is being profoundly jeopardized in scarcely a lifetime or two.

Political Science then may be seen as a profession of the study of the politics and the politics of the study of a profession. The glory and grandeur of its distant origins in an emerging profession of dialecticians, rhetoricians, and expositors when the fundament concepts of political life were constructed and explored demands that we contemporary political scientists not only be ever mindful of our discipline as a profession and its politics as a profession, but that we also acknowledge our responsibilities as knowers and actors within the politics we study. If not Plato, then surely Aristotle would be proud of us for that remembrance and that acknowledgment.

Chapter 7: Why Study Political Science? – Professor Bruce Lapenson, PhD

Bruce Lapenson, PhD
Associate Professor of Political Science
North Carolina Central University

Dr. Bruce P. Lapenson received his Ph.D. in Political Science from Temple University and is Associate Professor of Political Science, specializing in political and social theory at North Carolina Central University. Dr. Lapenson's teaching interests are mainly in political theory, though he also teaches in the American Government sub-field. He has also taught intellectual history. His research interests are in political ambiguity in Freud, democratic theory and public policy, and race and philosophy. Dr. Lapenson is a former psychotherapist and social service supervisor also. He lived most of his life in Philadelphia, Pennsylvania, and currently resides in Durham, North Carolina. He went to high school in Trenton, New Jersey, where he was on the wrestling team. Dr. Lapenson's non-work related interests are sports, eclectic reading, film, art and music history, and exercise. He also loves cafes, and has great friends in Reverend Derric Jackson of Philadelphia, and Angelique Justin of Colorado Springs, Colorado, in addition to his political science colleagues at North Carolina Central University. Dr. Lapenson's father, Jerry Lapenson resides in Boynton Beach, Florida and recently retired from amateur tennis.

The Big Questions

The academic discipline of political science asks the big questions: What is government? Why is government necessary? What is freedom? Can someone truly be free living under government? What are rights? What is meant by citizenship? What can government ensure? The responses to these questions are as varied as the weather. Linking the questions of freedom and government, we can ask: Does government restrict individual freedom or provide the only possible environment for it? Austrian psychoanalyst and social philosopher, Sigmund Freud, proclaimed:

The liberty of the individual is no gift of civilization. It was greatest before there was any civilization, though...it had...no value, since the individual was scarcely in position to defend it.[i]

To Freud, freedom without government was no freedom at all. Life without government, even to those philosophers with relatively positive accounts of human nature, such as John Locke, imagined that humans created government because life without government (the state of nature)[ii] became inconvenient.[iii] Anarchist Emma Goldman however, challenges the necessity and good of government:

It matters not whether it is government by divine right or majority rule. In every instance its goal is the subordination of the individual. [iv]

To Goldman, negative human behavior was fostered by law and authority since it suppressed humans' instinctual need for freedom and self-expression. French philosopher and forefather of contemporary commanitarians,[v] Jean-Jacques Rousseau adds a third opinion to our present discussion. Rousseau theorizes that true freedom is not natural freedom, the freedom we have in a state of nature, but civil freedom, which is possible only when citizens of a community put aside their individual interest and form a general will. A general will is made up of what will benefit the whole political society in the long term, in short, it is rational political thinking, possible only when we refrain from giving into our individual appetites and begin to focus on the common good.

For Rousseau, government in itself was not automatically good. Liberal-democracies, for example created the conditions for appetitive natures and self-centered interest groups, (partial governments) to dominate law and policy. Natural freedom could therefore thrive under government, but would engender social and political conflict to the point where the citizen is not free. The only true freedom to Rousseau is civil freedom, in which the common good informs public decision-making.[vi]

Rousseau takes the questions we posed about freedom and its relationship to government to questions of citizenship. What does it mean to be a citizen? A simple definition of citizenship allows for minimal requirements i.e. one is a citizen per having fulfilled the legal criteria. Other depictions are more demanding. Aristotle believed that one had to earn citizenship, which meant the opportunity to participate in political debate and to hold political office. Aristotle viewed such activity, as expanding the intellect, which he believed required leisure time to accomplish.[vii] An American sociologist stated recently that some Americans were not ready for citizenship, based on, as he saw it, their lack of individual responsibility.

The Focus on Values and "Shared First Principles"

The focus on values and "shared first principles"[viii] as foundations of a community of citizens, achieved through participation in political and social affairs asks much more of each individual than does the mere legal criteria of what constitutes a citizen. These two notions of citizenships and what constitutes a community pose additional questions: Which view is best? If "citizen" is merely a legal term, what is our relationship to one another? What do we owe to each other, beyond civility?

On the other side, requiring community involvement, imposes on an individual's freedom to choose to do so, and once there is agreement on foundational principles, what of those who stand outside the agreement, the new table of virtues? At what point does liberal and democratic tolerance for varying opinions become threatened?

The broad and abstract notions we have been considering relate to more concrete questions: What is the proper role of government and what should it ensure or be responsible for? Again, we encounter disagreement. Harold Lasswell said that government decides who gets what, when, and how. [ix]Govern ment therefore, is a distributor of various public goods, bridges, parks, water treatment plants, but also taxes and determines what degrees of liberty it thinks are appropriate as regards various actions of citizens.

Does our liberty extend to burning the US. Flag or wearing a shirt made of an actual flag? A decision of the U.S. Supreme Court, regarding flag-burning controversies[x], is government deciding who gets what, when, and how. But, such concrete political matters cannot possibly be devoid of, or divorced from the abstract question we had considered. For example, on what underlying assumption about human nature, human needs, or individual freedom might a liberty to burn a beloved symbol as an expression of free speech or political protest be founded on? What underlying ideas inform those who oppose a legal right to burn the flag? Would such ideas be grounded on assumptions about the establishment and commitment to shared first principles?

Government Involves the "Authoritative Allocation of Values"

Another widely known view among political scientists concerning government role is contributed by David Easton. For Easton, government involves the "authoritative allocation of values"[xi]. Government provides values to its citizens though its decision-making, via laws, policies, and actions. While this definition of government's role sounds more abstract than Lasswell's, remember that the distribution of public goods, both tangible, funding for public works, and intangible freedom of speech, are founded on some value, something decided by government to be of value, and therefore encouraged. Stop signs and traffic lights speak to us, or more accurately, they become the voices of government promotion, or coercion, of something conceived as good.

So politics is more than "pork" i.e. politicians obtaining goods for their constituents. Politics is also about human behavior and social interactions. Liberal thinking, by some accounts, premises an "unencumbered self,"[xii] a self-committed to a value only voluntarily. Still, liberalism is promoting that very belief as a value. In other words, this particular interpretation of liberalism values individual decisions about values as its <u>fundamental value.</u>

Our discussion has brought us to a realization: Political events, laws, wars, foreign policy, cannot be separated from underlying beliefs. Thought informs actions. This is why political science sub-fields such as political theory are so vital to the other central areas of political science: American government, international politics, comparative politics, and constitutional law, which deal with the more concrete questions concerning practical political matters.

What isn't Political?

Returning to the overt world of laws, policies and government decision-making, we have to ask this question: What isn't political, or more specifically, what is not a product of law and policy or effected by law and policy? A stroll down your street reveals that the houses are subject to real estate taxes and building codes. The sidewalks are constructed by local government, cars regulated by state inspection and emissions standards. Your neighbor's job at an oil refinery requires that he wear specified protective gear as required by the federal government. You have stopped for breakfast at, Tim's Morning Spot, Tim had to get authorization from the city to build his store on land that he purchased, and he cooks food which at some point, prior to him doing so this morning, had to satisfy FDA (Food and Drug Administration) regulations. Government decisions permeate most areas of daily life, but lets look at the topic differently.

Besides laws and policies, where else do we see government influence? When law is silent[xiii] individual choice is unimpeded. In other words, if government has not spoken regarding a specific action, citizens may act or not act by choice. There is no law establishing a list of professions that I must choose from when I consider what I want to become. So, even when government is silent, it influences. Government could have created a list of career choices, but it chose not to. I was therefore influenced by its silence to make a personal decision.

Does politics influence behavior? Aristotle made the link between constitutions, regimes, laws, and actions. A constitution can be democratic or not so democratic. The political philosophy inspiring a constitution is responsible for the type of government that is created. A democratic constitution produces a democratic government, which as a result makes democratic laws, which, in turn, influences the thoughts and actions of citizens[xiv].

For example, participating in a liberal-democratic society promotes respect among citizens for opinions they do not adhere to, or even, find despicable. So, even highly unpopular organizations, in the US, have the right, because of the 1st Amendment in the US Constitution to express their viewpoints in public. Since all individuals, organizations, and groups hold this right, a democracy emerges, in which the right of such expression is joined with the right of a counter demonstration. This is all made possible, of course, by a liberal-democratic constitution.

Careers: the Political Science Degree can Lead to Opportunities in Various Occupations

So far, we have focused on the relevance of the study of political science to the institutions which influence our lives, the ideas which inform such institutions, and from which we are influenced. We turn now to the most pragmatic of concerns: Careers. The political science degree can lead to opportunities in various occupations. North Carolina Central University political science graduates have went on to Law School, Masters, and Ph.D. programs. Some are currently working as college professors, lawyers, administrators, US. Congressmen and congressional staff members, political campaign staff, pursued military options, and civil service positions.

The B.S. in political science has also functioned as a general liberal arts degree, viewed by employers as providing an eclectic, well-rounded education to graduates, and thus qualifying them for positions in numerous fields, education, retail, business, and health care are some of the fields.

Some of our graduates have also been accepted to Medical School and many of our graduates have used POLS 4100, an internship course, as a springboard for career opportunities. Others have taken advantage of internships we inform them of, and through which they establish contacts and pursue career opportunities. As one can ascertain, the study of political science has many educational and practical rewards.

Chapter 8: Now is an Especially Exciting Time to Study Political Science – Professor Diana Owen, PhD

Diana Owen, PhD
Associate Professor of Political Science
Georgetown University

Dr Diana Owen is Associate Professor of Political Science and teaches in the Communication, Culture, and Technology graduate program, and has served as Director of the American Studies Program. She is the author of ***Media Messages in American Presidential Elections*** *(Greenwood, 1991),* ***New Media and American Politics*** *(with Richard Davis, Oxford, 1998), and* ***American Government and Politics in the Information Age*** *(with David Paletz and Timothy Cook, 2012). She is the co-editor of The Internet and Politics: Citizens, Voters, and Activists (with Sarah Oates and Rachel Gibson, Routledge, 2006) and Making a Difference: The Internet and Elections in Comparative Perspective (with Richard Davis, Stephen Ward, and David Taras, Lexington, 2009). She is the author of numerous journal articles and book chapters in the fields of civic education and engagement, media and politics, political socialization, elections and voting behavior, and political psychology/sociology. She has conducted studies funded by the Pew Charitable Trusts, the Center for Civic Education, and other sources. Her current research explores the relationship between civic education and political engagement over the life course and new media's role in politics.*

My Story

I entered the George Washington University as a freshman fully intent on pursuing a career as an attorney. My family had been very involved in local politics, and I was interested in learning more about American government and how it worked. I enrolled as a political science major envisioning that this would be stop on the road to law school. My plans, however, changed radically in short order. I was captivated by the subjects I was studying in just my first semester. A course in public opinion and political socialization explored how people develop their identity as citizens from early childhood through their senior years.

Another class illuminated the central importance of the mass media, particularly television, in political life. I saw how candidates could manage their media strategies and employ political advertising to influence voters' perceptions during elections. A seminar on judicial politics convinced me that I would be happier studying the voting behavior of Supreme Court justices than practicing law.

Following the sage advice of my academic mentors, I headed to the University of Wisconsin-Madison to graduate school in political science where I focused on American government, political psychology and sociology, and research methodology. The decision to forego law school to study politics was the best decision I could have made.

As a political scientist, I have the opportunity to explore the things that interest and excite me the most through a political lens. Much of my teaching and research is tied directly to the fast evolving political media environment, especially the impact of social media on citizens, political leaders, government institutions, and journalists. I also work on the related topic of how to educate citizens to engage in politics in the new media era.

Use of Facebook, Twitter, Tumblr, YouTube, and Other Social Media in Politics

My research focuses on how civics instruction in elementary school, junior high, and high school prepares people to become active citizens over their life course. I am especially interested in the ways that schools are (or are not) preparing adolescents to use social media to take part in politics. In addition, I have the opportunity to teach classes on the intersection of sports and politics. Sports is often a hook that draws attention to wider political issues.

When I first began studying media and politics, television, print newspapers and magazines, and radio were the main sources of information. The influence of media in politics has increased significantly since new media entered the mix alongside traditional sources in the late 1980s beginning with rudimentary websites on the internet.

Although today many people still rely on television for political information, a revolution in the political uses of digital media is well underway. Social media, like Snap Chat and Vine, are digital platforms that allow people to connect, interact, and share content through diverse networks. It's fascinating to see how people's access to social media through a growing number of devices—from computers and tablets to cell phones—has changed their relationship to the political world. The public has new options for getting involved in politics that puts them on greater par with politicians and journalists. The political media environment is changing so quickly that there is always a novel development to study or an evolving trend to discuss in class.

A good deal of my research focuses on new media in election campaigns. The 2008 presidential campaign was a laboratory for examining the ways in which voters' innovation with social media helped to reshape electoral communication. Candidates from both the Republican and Democratic parties participated in "Facebook debates," and answered questions submitted by voters who created short video clips. Supporters of Democratic presidential candidate Barack Obama, especially, used social media to generate publicity and organize volunteers on behalf of his campaign. People no longer remained passively in the background watching while politicians and journalists on television took center stage.

Instead, citizens used Facebook, Twitter, Tumblr, YouTube, and other platforms to supply information, express opinions, and create online videos on behalf of candidates and issues they supported. Average people acting as "citizen journalists" captured live moments from the campaign trail, offered up-to-date accounts of events, and provided color commentary about candidates' visits to their home towns. "Citizen journalists" often scooped the professionals, and provided an alternative to official reports of campaign activities.

In more recent elections, campaigns have elaborated on these innovations. They also use "big data," archives of political and consumer information about individuals, to target voters with messages based on their preferences. Campaigns can create text messages designed to reach people with very specific profiles, for example a 25 year old graduate student who enjoys Starbucks coffee and science fiction films. However, it is unclear how successful this type of microtargeting is in influencing voter's opinions.

The Rise of Social Media has Changed Politics

The rise of social media has changed politics in other ways. Journalists are less likely to use the techniques of investigative journalism that take them into the field to track down stories, like Bob Woodward and Carl Bernstein employed when they exposed the Watergate scandal in the early 1970s.

Instead, they increasingly rely on information gleaned from websites, blogs, video sharing sites, and social media, including content provided by the public, to craft their stories. Social media also has become a means of mobilizing the public. The shooting of an unarmed teen by a police officer in Ferguson, Missouri, generated a social media firestorm that awakened dormant tensions over racial injustice and sparked protests across the nation. Politicians still use stump speeches and press releases, but they also take to social media to reach voters. They actively court Facebook friends, and make their case to the press and their constituents by tweeting 140 character sound bites that they hope will generate press attention, even controversy.

In addition, the role of humor and satire in politics has become more important. Social media enables widespread sharing of parodies and memes. Hundreds of memes have been generated based on a photo of Hillary Clinton using an outdated Blackberry device on an airplane and actor Clint Eastwood talking to an empty chair at the 2012 Republican National Convention. People not only enjoy the humorous content, they also learn from it. Fake news programs like *The Daily Show* with comedian John Stewart are especially popular with young people who tend to ignore traditional nightly newscasts.

Young people are playing a pivotal role in shaping the new political media environment. As digital natives who have grown up with modern-day technology, young people are in a unique position to shape the future of politics. They often are the first to develop novel political applications of social media. The importance of innovations devised by young voters in American presidential elections cannot be underestimated. They precipitated a fundamental change in the way that campaigns are waged by using social media to develop peer-to-peer networks to mobilize voters. At the same time, young people tend to be less involved in politics than older citizens, and a minority use social media to engage.

Social media provide opportunities for new forms of participation, but they also pose the practical challenge of educating the public to be savvy and responsible users. Civic education in schools is one possible mechanism for preparing young citizens to use new political media to take part effectively in politics. Integrating programs of digital political literacy in schools, however, is challenging, and many schools do not have the resources to incorporate it into the curriculum. Further, schools with at-risk students are the least likely to have digital political literacy programs, which works to widen the participation gap between rich and poor.

The Political Culture of Sports

Finally, I am an avid sport fan, and my political science background has helped me to better understand the integral connection between sports and politics. I have been able to integrate my love of both sports and politics into my job by teaching a course entitled, "The Political Culture of Sports."

The students and I delve into topics from an historical perspective, such as how politics plays heavily into the Olympic games and the evolution of college sports as big business. We tackle topics within the context of sports, including racial and gender discrimination. I find that almost daily media headlines demonstrate the relevance of sports to modern day political affairs. Presidential candidates routinely use sports metaphors to appeal to voters, and their athletic prowess becomes a measure of their fitness for governing.

Every year, President Barack Obama makes public his picks for the NCAA basketball tournament brackets to much fanfare, and it is perceived as a way of making him more relatable to the public. Sports can make visible difficult societal issues, and prompt calls for change. A video released by the tabloid TMZ of a football player dragging his wife from an elevator in Atlantic City after striking her during an argument sparked a national debate over domestic violence. The National Football League was forced to address the issue, and fund initiatives to deal with abuse.

Every day I find relevance of political happenings to my research and scholarship. I am most fortunate to be able to explore political trends as they unfold with students and colleagues. Now is an especially exciting time to study political science, especially with new phenomena unfolding, such as new media's influence on politics and government.

Chapter 9: Political Science: An Endless Song with a Variety of Rhythms - Dr. Kwame Badu Antwi-Boasiako

Kwame Badu Antwi-Boasiako, PhD
Associate Professor and Chair of Department of Government
Stephen F. Austin State University, Texas

Dr Kwame Badu Antwi-Boasiako is the Chair of the Department of Government and Associate Professor of public administration and political science at Stephen F. Austin State University, Nacogdoches, Texas, USA where he teaches in the Public Policy, Program Evaluation, Research Methods and American Government. He was a broadcast journalist (Ghana Broadcast Corporation) in Ghana from 1980-1992. He had his bachelor's degree in journalism (Broadcast News) from University of North Texas, USA, in 1996. He obtained two masters degrees in journalism and public administration in 1998 and 2000 respectively from the same university. He got his Ph.D (Public Administration and Policy with International Relations Minor) from Mississippi State University, Starkville, USA. His primary research interest includes traditional institutions and democracy in Africa, decentralization, terrorism, and diversity in the public sector. Some of his recent publications are: Rethinking utopianism, international relations theory, and marginalization of Africa in global politics. Journal of Comparative Politics. 7 (2): 34-51 (2014). And "The Trans-Atlantic Relationship: From Oppression to Hope of Partnerships." In Transatlantic relations and modern diplomacy: An interdisciplinary examination. Edited by Sudeshina Roy, Dana Cooper, and Brian Murphy, pp. 114- 124. Routledge: Taylor & Francis Group, New York, 2014.

Introduction

Any attempt on my part to define political science in one sentence would definitely jeopardise the true meaning of the discipline; like most academic disciplines, it comprises many subfields that could not be covered in this chapter. Here, a very brief attempt is made to discuss and define political science; however, it by no means covers everything political. There is a conundrum regarding whether or not political science is part of history or history is part of political science (Seeley, 1919).

According to Seeley there is a dance between history and political science; therefore, the two complement each other as academic disciplines. He states, "History without political science has no fruits; *and* Political science without history has no root" (4). He explains the relationship between the two and argues that a systematic understanding of the past is scientifically essential for human development and generational change. To him, political "science has for a long time past been invisible growing up by the side of history and everyone has perceived that it has some connections with history, and must draw a great part of materials from history" (13).

This inseparabilityof political science from other subjects does not only reside with history but could be argued for other subjects, including public administration. For example, the genesis of public administration as an academic discipline, which sees Woodrow Wilson (1887) as one of its pioneers, has an endless back and forth dance with political science (Henry, 2010). In his article, *The Study of Administration,* Wilson struggled to separate administration from politics. Unfortunately, for those who believe in the Wilsonian argument, that notion has not always materialized. In fact, in 1900, Frank Goodnow identified two distinct functions of government, which were *politics* and *administration*. To Goodnow, as cited in Henry (2010:28), *politics* "has to do with policies or expressions of the state will," while *administration* "has to do with the execution of these policies."

So the dance here is whether public administration is a subfield of political science or not. This argument and distinction is clearly beyond the scope of this chapter. However, the argument now is that political science as an academic discipline has many tentacles, which makes it difficult to define. So how do we define political science given the above confusion presented? This chapter defines political science, discusses why we study political science, and concludes with the future of the discipline regarding practitioners and scholars in the global village.

Defining Political Science

Political science, by and large, provides platforms for explanations of activities of government in world affairs. Government here refers broadly to a vast array of national, state, regional, local, municipal, city, village, and traditional institutions.

For example, in Britain, Norway, Ghana, Australia, and Italy, government at the national level may mean the president or prime minister and his or her cabinet under the parliamentary system. The same is true for the United States but government at the federal level applies to the president (executive), congress (law making) and judiciary (courts). There are other forms of government in the United States, which comprises the states, counties, municipalities, and cities.

In other less developed societies, the village government, which can reveal an astonishing functionality, may comprise the chief and council of elders to ensure "rule of customary law and accountability" (Ayittey, 2005:21). Ayittey argues that African societies and other indigenous tribes elsewhere had rulers who were "surrounded by various councils, bodies, and institutions to prevent abuses of power and corruption." Those types of governments in so-called "primitive" societies may not fit the scholarly definition and understanding of political scientists, but they are governments, since they have institutions and procedures through which their lands and people are ruled (Lowi et al., 2012: 5-7). Other known forms of government include autocracy, oligarchy, authoritarian, totalitarian, and constitutional.[xv] All these governmental structures are covered by political scientists in the academia, hence the conundrum.

The problem here is that, unlike the pure sciences where defining subjects such as physics, chemistry, and/or biology are precise and concise, the same argument cannot be made for the social sciences, including political science (Antwi-Boasiako, 2014:115 and Seeley, 1919:1-29).

In fact, Ethridge and Handelman (1994 and 2013) are more precise with the ambiguity and struggle political scientists encounter in defining their discipline compared to those in the pure sciences. They write, "Political scientists envy physicists when it comes to defining their basic concepts. The physics instructor encounters no preconceived notions among students when he or she speaks of the [coefficient of kinetic frictions,] ..." They maintain that "academic definitions of politics and government may strike us as abstract and sterile, perhaps because they are designed to distinguish between popular and scholarly uses of the terms, hence two definitions are particularly important" (6).

Politics over the decades are understood to have negative connotations as political appointments happen through the spoils system theory and connections. Though various political scientists have made attempts to define the discipline, they do not seem to settle on one definition. Below are a few definitions:

- "Politics is the science of who gets what, when, and how."[xvi]
- "Politics is the authoritative allocation of values." [xvii]
- "Politics is the processes by which human efforts towards attaining social goals are steered and coordinated."[xviii]
- "Politics is... the activity by which differing interests within a given unit of rule are conciliated by giving them a share in power in proportion to their importance to ... the whole community."[xix]

The above by no means exhausts the plethora of definitions provided in the literature. In fact, it is the tip of the iceberg. The American Political Science Association (APSA) has a shorter but inclusive definition, which states, "Political science is the study of government, public policies and political processes, systems, and political behaviour."[xx] The Association further provides a rather detailed list of subfields[xxi] of political science as an academic discipline, which even makes it more cumbersome to accept one definition for political science.

It is quite interesting to note that most people, especially the youth, are more likely to divorce themselves from politics, but their very existence and behaviour is politically *controlled,* whether one resides in any of the democratic countries such as Australia, Great Britain, Nigeria, Ghana, and the United States of America, or in other countries where different forms of governments other than democracy exist, such as Communist China and North Korea. If the definitions provided above are anything to go by, then the obvious question becomes, why do we study political science, since we are all political beings, including those who claim to be apolitical? Is political science of interest only to academia or do students and practitioners also need to know more about politics?

Your answer may be different from others, but it is that uniqueness in opinionwhich makes the study of political science interesting and worth studying. Before we jump to the next section, let us ask, why study political science? Charles Lindblom (1997) has a rather unique definition or understanding of political science. He maintains it "is a name given not to a field of conventional scientific inquiry but to a continuing debate."

This observation of Lindblom, I argue, stems from the belief that "on any given big issue of fact or value, debate in political science tends to be endless rather than declining" (260-61). The validity of his observation could be that if one looked at the issues of inequalities, immigration, and discrimination in the United States of America, for example, there has been a series of discussions over these issues, yet there seems to be no end in sight. In fact, no one knows when these issues would never be discussed again, hence the justification of Lindblom's definition of political science.

Katznelson and Milner (2002) also agree with Lindblom's observation: "the image of a long-running debate captures well the dynamics of our discipline as a continuing process of inquiry engaged with understanding fundamental political issues."(2). If political science is about long debates, as proposed here, why don't you study it, so that as a participant of any debate you must present your facts in a chronological fashion, instead of haphazardly presenting your opinion without evidence to support your argument? At this point we must also attempt to understand science before we dive into reasons to study political science.

Babbie (2001) sees science as a method of inquiry, which provides a way of learning and knowing things about the world around us. He opines science is "conscious, deliberate, and rigorous undertaking" but the author is quick to point out that scientists disagree "on proper definition" (15). To Monette, Sullivan and DeJong (2005), science is "a method of obtaining objective knowledge about the world through systematic observation" (22).

Note that there are five characteristics that are associated with science, according to the authors, and these are: empirical, systematic, search for causes, provisional, and objectivity (22-23). We should also remember that objectivity is subjective, hence the definition has ambiguities. The lack of consensus should rather encourage us to strive in identifying a more precise definition for political science.

Why Study Political Science?

This is my first day of political science class activity, before I even discuss my syllabus with my students. On the first day of a semester, I will always ask those who are *idiots*, a class of fifty (50), to sit in front of the classroom and astonishingly almost everyone would move to the back rows. Then I will ask again, those who are not interested in political science or politics must move and sit at the back of the class; then suddenly the very people who moved back would stream to the front seats. So why, with the same question, do I get opposite reactions by the same group? The simple test to new college students gives me the impetus to argue for why political science must be studied. There are a lot of us who claim to have political knowledge of the political system we reside in, but very soon one realizes how ignorant we are about our own political structures.

Political ignorance is a disease, which prevents us from actively and positively participating in the political system. Acquiring political knowledge is essential for progress, since knowledgeable citizens contribute in diverse ways toward national development (they pay their taxes, observe the laws and regulations of the land, vote in elections and other things required of a good citizen). As James Madison noted, "Knowledge will forever govern ignorance. And a people who mean to be their own governors must arm themselves with the power that knowledge gives."[xxii] Political ignorance is not a new phenomenon as Somin (2013) opines.

According to Somin, political philosophers have debated the pitfalls of voters' lack of political knowledge in a democratic society since the genesis of democracy in ancient Greece, in Athens. He goes on further to argue that democracies are "doomed to failure because its policies *are* set by ignorant common citizens" (7). The question you should ask yourself is, would democracy grow or fail globally? Political knowledge is defined here as awareness of factual matters related to governance, politics, and public policy.

To prevent the failure of democratic principles, political science as an academic discipline helps to educate the people to uphold democratic values and maintain a well-structured political system. As Somin noted, "many voters are unaware of the elements of important public policies enacted by the legislature" (9) and it is through the study of political science that political awareness would come to the fore.

So it becomes clear that political science is not only important for those in the academia but also to all of us. We must be aware of how rules and regulations are made, so political science is equally important to the scholar and the practitioner. Now let's get back to our first class day exercise and the meaning of *idiot* in political science. Those who study linguistics would agree that the meaning and use of certain words do change over time. Referring to one as an *idiot* is insulting, but the original meaning of this Greek word refers to anyone who shows no interest in politics or affairs of the state, your country.

So, even if you think you belong to the *idiots*, it must be binding upon all useful and progressive citizens to learn about the political systems in which they reside. As a scholar, you have to know something about the nature of government of your country and how other political systems function to have a better understanding of the international political environment (international relations).

Since reasons for studying political science have value in different ways, which may not be exhausted in this chapter, I would like to draw your attention to another reason why this discipline is important to you. Given the definitions provided above, it becomes clear that a degree in political science equips you to prepare for graduate school regardless of the major you decide to pursue. The reason is that it prepares students to become confident debaters (oral communication), become good research driven individuals (both qualitative and quantitative skills), and exhibit analytical reasoning (both written and oral communication). To those who might take an earlier career route, such a degree ensures confidence in you.

For example, if you decide to become a freelance broadcast journalist reporting from other parts of the world, your training in political science helps you to understand the types of governments and countries you might be reporting from. This means that you, as a reporter with a political science background, are more prepared in your analysis of political events in other countries.

Globalization, global politics, foreign policy, public policy, international political economy, international trade, international organization, international sports events, and many other global topics continue to justify the importance of political science. The purpose of political science as indicated earlier is multi-faceted; it has been established that studying political science helps the student and common citizens to gain improved understanding of world politics, while identifying current developments in international relations (Rhodes, 2006). It also helps to "explore issues that will require careful and painful decisions on the part of policy makers in the years ahead" (Spiegel, 1991:1). So here, Spiegel may be arguing that, as a student of political science, you are more likely to develop better understanding of political activities and ideas that shape future policies. Another way of understanding political science is through theories, which are integral part of studying social sciences.

Theories are of great importance to the discipline. As Babbie (2001:51) opines, they "are systematic sets of interrelated statements intended to explain some aspect of social life." Thus, the logical starting point is formulating, a series of linked propositions derived from a chronological collection of observations on a particular aspect of life. It must be kept in mind, however, that "no one theory will ever be sufficient to yield a complete explanation to a phenomenon in the social sciences" (Antwi-Boasiako, 2014b: 41). Theories are developed or formed in political science to explain a given phenomenon but the formulation of theory, as Dubin (1978, 6) sees it, "lies in the human behavior of wanting to impose order and unordered experience, which is not ordered by nature hence the experiences may be...theorized about, in very different ways." Put differently, the amalgamation of different variables must be incorporated in theory construction hence the need to examine different variables in theory formulation in political science.

Career and Practitioners

Unlike public administration, criminal justice, nursing, and other academic disciplines that have a clear and distinct career path, well established in the world outside the academia, political science majors might be sceptical regarding their career choice. However, unlike those majors indicated above, studying political science does not box you into one field. Instead, it provides tons of opportunities for people who plan to work in the public or private sectors, including non-profit organizations. For example, one can plan to follow a career path in government as a social advocate, a lawyer, or head of a diplomatic unit. In the era of democratic ideas and principles waving across the globe, political parties are looking for qualified individuals who have better understanding of the politics within their states, counties, municipalities, and cities to be part of their campaign teams. Multinational corporations, private businesses, non-governmental, and governmental organizations do not only look for individuals who speak multiple languages but, as employers continue to establish businesses in other countries, they also prefer people who have both political knowledge and awareness of other cultures.

Benefits of Studying Political Science at a Mid-size University

By definition Stephen F. Austin State University (SFA: www.sfasu.edu) is considered a mid-size university in the United States of America with a population of nearly thirteen (13) thousand students (http://www.sfasu.edu/research/docs/FastFacts.pdf.). It is a regional university locate in Nacogdoches (the oldest town in Texas); East Texas but has both out of state and international students (http://www.sfasu.edu/oip/). This means there is cultural and international diversity at SFA. Our faculty comprises well qualified American and international professors who are experts in their fields of study. Given this brief background, the uniqueness of studying political science at SFA include but not limited to small class sizes, personal (professor-student) interaction, professors guiding their students into the future, and placement into law schools, graduate schools, and both private and public sectors.

According to the University's Office of Institutional Research, the teaching faculty-students ratio for the Fall Semester 2013 was 1:19 University wide. Benefits for studying at a mid-size university in America such as SFA, are the availability of scholarships for both American and international students (http://www.sfasu.edu/oip/210.asp). Our dedicated political science professors have diverse research interests and background and their publications range from political theory, political history, political thought, human rights studies, American government, presidency, international relations, terrorism, budgeting, and local government administration. The complexities and ambiguities in the study of political science are seen in the discipline's "inability to identify a common methodology, a common approach to the study of politics" (Scott and Garrison, 2002:13).

Despite that difficulty, our faculty work very closely with our students because of small size classes. Writing a research paper in political science is essential in understanding the discipline (Baglione, 2012). Our faculty focus on writings because effective communication establishes understanding among all parties. As Scott and Garrison noted the works of political thinkers and philosophers are read today not "only because of the quality of their ideas but also because of their ability to communicate clearly and powerfully (Scott and Garrison, 2002:13). The resultant effect is that faculty make sure that student know the differences between plagiarism and paraphrasing, how to identify researchable topics, methodology, design of students papers and how to connect the various parts of a paper together.

Students also benefit from faculty collaborative research work as faculty members do publish with their peers in other disciplines. For example, some of my publications are in collaboration with other professors from geography and communication departments while others from other universities or in another country. [xxiii]

The Future of Political Science and Conclusion

Discussing the future of political science is not a new phenomenon. In 1963, Harold Lasswell authored a book, *The Future of Political Science*, and looked at the past, present, and future of the discipline in detail. In his analysis, Lasswell did not only discuss the future but also focused on some of the challenges and trends in the past. Such detailed analysis, which is not done here in this chapter, helps one to follow the trend of growth in terms of the number of students getting their first degrees, masters, and Ph.D.s in political science. It should be noted that the prospects of political science to grow as a discipline, as seen by Lasswell, are true today.[xxiv]

Given the divergent meaning of what comprises political science, there are now degrees in international relations, international affairs, public administration, political history, women and gender studies, western political thought, and many concentrations of political science. The future is very bright for the discipline as, in addition to the traditional political journals, there are now online political journal and blogs that help to spread the “gospel” according to political science, concerning government activities, functions, and policies. More and more people, apart from seeking careers within their various governments, are also going into teaching, research, and consultancy. This further affirms the growing importance of political science to scholars and practitioners.

In this chapter we attempted to define political science by looking at the literature. Political science is seen as a discipline with many associations. In fact, the various concentrations within the discipline include but are not limited to, international political economy, international relations, public administration and many more tracks in political science. The various definitions presented in this chapter should encourage you as a student to be able to define political science the way you understand it. But your definition must have basis for verification and not just your opinion. We also looked at why we should study political science. The general consensus is that political ignorance may help voters to elect "wrong" people as their leaders. Both James Madison and Somin emphasized the power of political knowledge, which is a weapon for political empowerment. Hence, the study of political science equips us in our research, written and oral communication, and helps us to understand the world around us.

References

Antwi-Boasiako, Kwame B. (2014). The transatlantic relationship: From oppression to hope of partnerships." In *Transatlantic relations and modern diplomacy: An interdisciplinary examination.* Edited by Sudeshina Roy, Dana Cooper, and Brian Murphy, pp. 114- 124. New York, Routledge: Taylor & Francis Group.

Antwi-Boasiako, Kwame B. (2014b). Rethinking utopianism, international relations theory, and marginalisation of Africa in global politics. *Journal of Comparative Politics*, 7 (2): 34-51.

Ayittey, George B.N. (2005). *Africa unchained: The blue print for Africa's Future.* New York. Palgrave, Macmillan.

Babbie, Earl (2001). *The practice of social research 9th ed.* Belmont, CA, Wadsworth.

Baglione, Lisa A. (2012). Writing a research paper in political science: A practical guide to inquiry, structure, and methods 2nd Washington DC. CQ Press.

Dubin, Robert. 1978. Theory Building. New York: The Free.
Ethridge, Marcus E. and Handelman, Howard (1994). *Politics and a changing world: A comparative introduction to political science.* New York. St Martin's Press.

Ethridge, Marcus E. and Handelman, Howard (2013). Politics and a changing world: A comparative introduction to political science. United State, Wadsworth.

Henry, Nicholas (2010). *Public administration and public affairs*. New York, Longman.

Katznelson, Ira and Helen V. Milner (2002). "American political science: The discipline's state and the state of the discipline." In *Political science: State of the discipline*. Eds Ira Katznelson and Helen V. Milner, 1-26. New York, W. W. Norton & Company.

Lasswell, Harold (1963). *The future of political science*. New York. Atherton Press.

Lindblom, Charles E. (1997). "Political Science in the 1940s and 1950s." In *American academic culture in transformation: Fifty years, four disciplines*, ed. Thomas Bender and Carl E. Schorske, 244-70. Princeton, NJ: Princeton University Press.

Lowi, Theodore J. (2012). American government: Power and purpose. New York, N.W. Norton.

Monette, Duane, R., Thomas J. Sullivan, Cornell R. DeJong (2005). *Applied social research: A tool for the human services*. Belmont, CA, Thomason.
Scott, Gregory M. and Stphen M. Garrison (2002). *The political science student. Writer's Manual 4th* . New Jersey, Prentice Hall.

Seeley, Sir J. R. (1919). *Introduction to political science: Two series of lectures*. Macmillan and Co, Limited. St. Martin's Street, London.

Somin, Ilya (2013). Democracy and political ignorance: Why smaller government is smarter. Stanford, CA: Stanford University Press.

Spiegel, Steven L. (1991). *At issue: Politics in the world arena*. 6th ed. New York, St. Martin's Press.

Woodrow, Wilson (1887). The study of administration. *Political Science Quarterly* 2: 197-222.

Chapter 10: Politics: Fascinating, Factual and Fun – Professor Jacqueline Briggs, PhD

Jacqueline Briggs, PhD
Head of the School of Social and Political Sciences
University of Lincoln

Dr Jacqueline Briggs is the Head of the School of Social and Political Sciences at the University of Lincoln, UK and the former Vice Chair of the Political Studies Association of the United Kingdom (2011-14). She has two books coming out later in 2014, one entitled ***Doing Politics*** *(Routledge) and the other entitled* ***Teen Players: Young People and Political Participation*** *(Palgrave Macmillan).*

Why Choosing to Study Politics is a Clever Move!

So, you have chosen to study political science or perhaps you are contemplating studying the discipline in the near future? If it is the former, let me be the first to congratulate you on your excellent choice. If the latter is true then, hopefully, this short piece will help to convince you to study the subject that is at the heart of everything we do. The Ancient Greek philosopher, Aristotle, writing centuries ago, dubbed politics the 'master science' and that remains relevant today. The importance of politics should not be underestimated. There are many definitions of what we mean by politics.

Firstly, some say it is about the study of institutions (government, political parties, the role of the media, electoral systems, for example). Secondly, others say politics involves conflict, so that wherever a situation of conflict occurs then politics comes into play. This is because politics often involves difficult decisions about the sharing of scarce resources. In the words of one famous political scientist, Harold Lasswell, it's about who gets 'what, when and how'.

Clearly, if we decide to spend more on health care, we will have less to spend on education, for example. There will be winners and losers and politics involves sorting out those disagreements. Thirdly, a much wider definition is that politics is about everything. Everything we do involves politics as it concerns power. Using this approach, even deciding who does the dishes at home could be seen as politics as it may relate to the concept of power. Under this wider definition, sexual politics and office politics all enter the equation (see Adrian Leftwich's work for a detailed and fascinating discussion of what is meant by politics; likewise Colin Hay's book on why we 'hate' politics). The feminist slogan which states that 'the personal is political' captures this concept perfectly. Politics, therefore, *is* about politicians, it *is* about political parties and governments but it is also about decision making, about who has power, and about what goes on behind closed doors.

Young People and Politics

Having studied and taught politics for more than thirty years, I remain fascinated by why some people are passionate about politics and why others have not yet caught the 'bug'. The concept of politicisation is an interesting one. If you could bottle it, I am sure all political parties would take an interest. One area of political science that is worthy of detailed investigation is young people and politics – which relates to this concept of politicisation.

Compared with other sectors of society, in many countries, young people are much less likely to vote and participate in politics. Countries such as Australia and Belgium, for example, have compulsory voting and, therefore, this means that young people in these countries are more likely to participate in mainstream politics (see Briggs and Celis, 2010). It is interesting to investigate how engaged and interested in politics young people are in these countries.

In other countries, the United Kingdom being one, young people are much less likely to take part in politics at a most basic level of voting. At the last General Election held in the United Kingdom (in 2010) only 44 per cent of the youngest sector of voters, the 18-24 year olds, actually bothered to vote. If you break this figure down in terms of gender, young women are the least likely to vote – with only 39 per cent of 18-24 year old women voting in the General Election (compared with 50 per cent of 18-24 year old men).

Some commentators say that young people are not necessarily disinterested in politics but that they are less passionate about mainstream politics. Many young people, it is claimed, are more interested in single-issue campaigns and in areas like animal rights or environmentalism. It remains the case that, as older people are much more likely to turn out to cast their vote, it makes sense for politicians to focus upon policies aimed at the older generations. Free bus passes, free TV licenses for the over-75s, winter fuel allowances, just a few examples of policies aimed at older people – no coincidence perhaps that these are precisely the people more likely to turn out and vote for our politicians.

In addition, the average age of a Member of Parliament in the United Kingdom currently stands at 50 years of age. Can a fifty year old really adequately represent the needs and wishes of younger people? Surely, there should be more young people amongst our elected representatives? This is not to say that our representative bodies should be an exact microcosm or mirror image of society at large but there should be more of an attempt to actually be representative. There is a growing literature in the area of young people and politics.

At What Age People Should be Allowed to Vote?

One key topic is in relation to whether or not the vote should be lowered to sixteen years of age. In some countries this has already taken place, in Austria for example. Likewise in Scotland, sixteen and seventeen year olds will be able to vote in the referendum on Scottish independence due to take place on the 18th September 2014. It will be interesting to see whether these newly enfranchised voters do turn out to cast their vote and also which way they choose to vote.

More generally, there are arguments on either side in relation to whether or not the vote should be lowered to sixteen. Some people feel that it should be lowered given that sixteen year olds can pay taxes, can get married or serve in the armed forces. In addition, some say that it is a question of 'rights', that regardless of whether or not they will actually turn out to vote but that at least they should have the opportunity to do so. Those opposed to lowering the vote argue that many sixteen and seventeen year olds are not sufficiently mature enough to cast their vote, that many do not actually want to be able to vote, that even very young children pay taxes on items bought with their pocket money (so that the taxation argument is not relevant) or the thin end of the wedge argument, i.e. where do you draw the line?

Should ten year olds be allowed to vote? Eight year olds? Clearly, there are many arguments on both sides of the debate. Personally, I am a strong supporter of lowering the vote to sixteen and seventeen year olds (see Briggs, 2010). I believe that it will happen too. I also believe, however, that political education is crucial so that young people have a better understanding of what politics entails and how political systems operate. This is not to say indoctrination but simply teaching young people about the 'facts', about how laws are made, how the electoral systems works, who are our elected representatives, and so on.

You have to 'Agree to Disagree'

In summary, politics is fascinating, it is factual and it is also fun. Heated classroom debates may mean that you have to 'agree to disagree'. Topics can be brain-aching, such as is human nature basically good or basically bad? Do we need lots of rules, laws and regulations almost to protect us from ourselves? Should capital punishment be allowed, should a state be able to take life as the ultimate form of punishment? Should gay marriage be allowed? Should sixteen year olds be able to vote? Should quotas be introduced to get more women into politics or into the boardroom? The beauty of political science is that it is interesting to play devil's advocate or to try to see the other side of the argument.

In addition, a degree in politics can be passport to a whole host of fascinating careers. Some, such as political journalist, politician, civil servant, political researcher, are closely linked to the subject, others are less so but attainable by virtue of the fact that you are a political science graduate. Integrated internships, placements and study periods overseas all add to the value of a political science degree.

As stated, this ancient discipline remains relevant in the twenty-first century. The use of social media, such as Facebook™ and Twitter™, means that politics is immediate, relevant and contemporary. Globalisation adds to the subject matter. The shrinking world means that it is easy to study and investigate the political systems of other countries. Studying events and phenomena, such as the 2011 Arab Spring, the global economic downturn, higher education tuition fees, the lack of trust in elected politicians, youth unemployment, instil renewed interest in the discipline.

New and exciting ways of studying political science mean that not only is the subject matter relevant and topical but the way that it is studied makes it dynamic and fun. Numbers studying politics remain buoyant, illustrating its popularity as an undergraduate subject. If you haven't already made the decision to study political science as your major, hopefully, some of the comments above will enable you to make the right choice. Believe me, you won't regret it!

References

Briggs, J. E. (2010) *Generation i-Pod: from apathy to engagement*, e.Pol, Volume 3, Number 2, April, Manchester, Political Education Forum: 3-5.

Briggs, J. E. and Celis, K. (2010) 'For or Against: Compulsory Voting in Britain and Belgium', *Social and Public Policy Review*, Vol. 4. No. 1: 1-33.

Hay, C. (2007) *Why We Hate Politics*, Cambridge, Polity Press.

Lasswell, H. (1958) *Politics: Who gets What, When, How*, New York, Meridian Books, first published 1936.

Leftwich, A. (ed.) (2004) *What is Politics?*, Cambridge, Polity Press.

Chapter 11: Political Science and Its Study at Marquette University – Professor Lowell W. Barrington, PhD

Lowell W. Barrington, PhD
Department of Political Science
Marquette University

Dr Lowell Barrington is Associate Professor and Chair in the Department of Political Science. His teaching and research specializations include post-Communist politics, ethnicity and nationalism, democratization, and political science research methods. Prof. Barrington has received research grants from the National Science Foundation (NSF) and the National Council for Eurasian and East European Research (NCEEER). He is a Research Affiliate with UW-Madison's Center for Russia, East Europe, and Central Asia (CREECA). He has also served as Editor-in-Chief of Analysis of Current Events, Social Science Proposal Reviewer for the National Research Council, and was Vice President of the Association for the Study of Nationalities (ASN). He is currently a member of ASN's Executive Committee. Prof. Barrington's publications include an introductory comparative politics textbook, ***Comparative Politics: Structures and Choices****, currently in its second edition (Cengage, 2013). He also has an edited book (University of Michigan Press, 2006) titled* ***After Independence: Making and Protecting the Nation in Postcolonial and Postcommunist States****. His research articles have appeared in Comparative Political Studies, Europe-Asia Studies, European Journal of Political Research, Georgetown Immigration Law Journal, Harvard International Journal of Press/Politics, Journal of Central Asian Studies, Nationalism and Ethnic Politics, Nationalities Papers, Political Research Quarterly, Post-Soviet Affairs, Post-Soviet Geography and Economics, PS: Political Science and Politics, Review of Central and East European Law, and World Politics. He is also the author of* *book chapters on Baltic citizenship and nationalism and regional divisions in Ukraine.*

What is Political Science?

Political science is the systematic study of politics, including both international relations and domestic politics in countries around the world. To understand everything from protests, revolutions, coups, peaceful regime transitions, elections, and significant policy reforms to wars, terrorist attacks, international treaties, and humanitarian interventions can be a real challenge. Political science students learn about major concepts and theories that allow them to organize and make sense of the many causes and consequences of these kinds of political outcomes. They also study decision-making and details about the people – leaders and members of the general public alike – who make political decisions. Understanding such decisions also requires political science students and scholars to pay careful attention to the social, economic, and political structures in which political decisions take place.

Focusing on the role of individuals and the settings in which individual and collective political activities occur allows students to tackle the ultimate goal of the study of political science: developing the ability to understand and explain major political outcomes. Even when a political event or policy result takes place in a country students do not know much about, political science gives them the tools to know what questions to ask, where to look to get a sense of the important information they need, and what existing theories to consider as they work to understand that outcome.

What Are the Career Opportunities Like for Political Science Majors?

Because politics touches on so many aspects of economic activity, social relations, and daily life, developing the ability to understand it opens doors to careers in the wide-ranging areas of government, law, business, advocacy, and service just to name a few. We often say that it is easier to list the jobs you *cannot* do with a political science degree than all of the ones you can. As we look ahead, our increasingly multicultural society and ever more interconnected world mean that politics will only continue to grow in importance.

Undergraduates also benefit from political science's focus on research and writing skills and critical thinking abilities. Learning to assess and take apart a problem, to weigh alternative explanations and prescriptions, and to put together your own answer to that problem gives students a set of skills that can be applied to many occupations. Indeed, a number of our undergraduate majors and graduate students go on to work in fields that only somewhat touch on politics. These include positions in banking, media, communications, and research.

At the graduate level, whether through a master's program or Ph.D. studies, students get to take their research, analysis, and writing abilities to a higher level. They investigate interesting questions with answers that affect pressing social issues, and they conduct scholarly research based on and the collection and analysis of primary source data. Although many who earn their Ph.D. in political science go on to teach at universities, the research skills developed at the graduate level make such individuals increasingly attractive to important research institutes, think tanks, and advocacy organizations.

Practitioners and Political Science

It is often said that practitioners (non-academics/industry professionals) are generally not very interested in taking advantage of academic research in Political Science. This is a real problem, and it poses important challenges and opportunities to both practitioners and to political scientists. It is true that, sometimes at least, political scientists conduct research that seems to lack relevance to those practitioners working in the "front lines" of policy creation. For a long time, the American Political Science Association has been encouraging its members to engage in research that is more "policy relevant." Many of the best political science researchers have heeded this call, and there has been improvement in this area. Political scientists are increasingly involved in collaborative work with government officials, community service associations, the business sector, and international organizations.

It is also true that, without trying too hard to read research articles and scholarly books by leading political scientists, many practitioners have assumed political science research is irrelevant to their work. I find that strange. In nearly every profession, workers are encouraged to learn about their field, what works and doesn't work, and how it might be improved. Those working in the arena of politics can certainly benefit from the research of those who study and teach about politics for a living.

Political Science at Marquette University

The faculty of Marquette's Department of Political Science work on a wide variety of topics and study all parts of the world. We have a number of professors interested in political economy (the intersection of politics and economics), while many others focus on important public policy debates on topics that include urban life, national and global health care, welfare state protections, and the treatment of minorities and other socially marginalized groups. Several of our professors study the United States, seeking a deeper and often historically informed understanding of the development of modern American politics. We also have a strong commitment to political philosophy – the focused reflection on normative questions of how best to organize society, craft political institutions, and use (or limit) political power. Throughout all of these endeavors, we work to connect our scholarship to the practical concerns of how contemporary politics shapes the lives of those around us.

Examples of My Research and Interesting Findings

Like many people who work in the part of political science known as comparative politics, I have both thematic interests (democratization, nationalism, and research methods) and a regional interest (the post-Communist region, particularly the Baltic States and Ukraine). I look for topics and specific research questions that fall at the intersection of my thematic interests and allow me to pursue the answers to these questions by examining the post-Communist cases. Much of my initial research, for example, examined the citizenship policies of the Baltic States and Ukraine. I sought to understand the variation in how inclusive the policies were – i.e., how easy or difficult it was for ethnic minorities to become citizens.

These studies pointed to the importance of identity-related factors, such as the extent to which the state was seen as a nation-state and the nation's membership boundaries were understood to coincide with the membership boundaries of the majority ethnic group. One of the surprising findings from this research was how effective European international organizations were at pressuring countries like Estonia and Latvia to make their policies more inclusive, while pressure from powerful neighbors like Russia tended to have a negative effect if any at all. I have also studied the impact of regional divides within mass attitudes in countries like Ukraine. This kind of research can provide a useful context for trying to understand topics like the ongoing conflict between Russia and Ukraine.

Scholarship Opportunities for Domestic & International Students at Marquette University

Marquette University works hard to provide as much financial support to its domestic and international students as possible. It also makes bringing international students to campus a priority. We have a very good Office of International Education, whose staff works closely with international students to facilitate their transition to university life. There are also many scholarship opportunities for undergraduate and graduate students from other countries. As a result, our university is home to hundreds of international students from around 70 countries. Historically, 90% of our international students have received scholarship support.

At the graduate level, financial aid offers mostly come from the individual graduate programs themselves. Both domestic and international graduate students can apply for research assistantships (which include partial or full tuition coverage and a stipend) or tuition scholarships. We have had a number of very good international students in our programs over the last several decades, many of whom held research assistantships. We value their perspectives, experiences, and approaches to the topics we study in our graduate programs. Our students from the United States, and our many professors who work closely with the students in our graduate programs, benefit from the presence of international students in our program.

Final Thoughts

At good universities in the United States, political science students are pushed to go beyond memorizing facts. Learning the political, social, and economic details about different countries is interesting and allows a student to better understand those countries. It is even more important, however, for students to ask good questions and attack the process of answering those questions through careful study and a commitment to developing critical thinking skills. If students only learn facts, they have not developed the ability to apply those facts. Even worse, if students acquire factual knowledge that merely confirms what they already believe, it is doubtful they have learned how to answer the truly important questions about politics, economics, and society. The best political science students – and the ones who are the most successful at whatever they choose to do after college – are those who listen to, take seriously, and really try to understand the arguments of people with whom they disagree. Helping students reach that point is one of the most rewarding parts of what I do as a political science professor.

Chapter 12: Politics is the Master Science – Professor Robert Garner, PhD

Robert Garner
Professor, Department of Political Science and International Relations.
University of Leicester, UK

Dr Robert Garner is the Professor at the Department of Political Science and International Relations. The focus of his principal research interest is the political representation of non-human interests, where non-human interests are taken to include those of non-human animals and other non-sentient parts of nature. This is an area of study that fits into the broader subject area of environmental politics. Professor garner has recently finished a project, initially facilitated by a Leverhulme Trust Fellowship, on the relationship between theories of justice and the protection of animals. His book, A Theory of Justice for Animals, published by Oxford University Press, seeks to consider the value of considering the treatment of animals within the prism of justice rather than what I take to be the broader arena of ethics and morality. It is also the first work that attempts to apply both ideal and nonideal theory to animal ethics and, in particular, to develop a viable nonideal theory of justice for animals. Professor Garner is willing to supervise projects on environmental politics, broadly conceived, animal rights, pressure groups and social movements, and political ideologies.

What is Politics?

It is no accident that Aristotle, the great Greek philosopher, described Politics as the master science. It is preeminent because politics is an inevitable feature of all societies. We cannot do without it. Thus, a popular definition of politics is that it is the process by which groups representing divergent interests and values make collective decisions. Politics is inevitable because of three features of the human condition. First, we have to find a way to live together. We cannot exist in isolation from others. Second, all societies of any complexity contain a range of different interests and values.

Therefore, there will always be a need for a mechanism whereby these different interests and values are reconciled. This is politics. Third, scarcity is also an inevitable characteristic of all societies. Since there is not enough of the goods that people want to go around, there needs to be some mechanism whereby these goods can be distributed.

The study of politics prior to the nineteenth century was almost exclusively concerned with a study of values. That is, politics was equated with philosophy. Political philosophers asked what is the good life? What, in other words, is the best kind of society for us to live in? Political philosophy is still an important component of the study of politics. It asks normative, or ought-type questions.

Nowadays, though, political philosophy is only one branch of the politics curriculum. It is accompanied by empirical political science, on the one hand, and international relations or politics, on the other. The former asks empirical, or is-type questions about political phenomena. These include the study of particular political systems, sometimes individually, and often in a comparative framework. The study of international politics or international relations has become increasingly popular in recent years. The relationship between sovereign states has been an important component of the study of politics for decades.

More recently, however, the focus of politics has begun to shift because in a practical sense we are living in a world which is becoming increasingly interdependent, where the forces of so-called globalisation are placing increasing constraints on what individual so-called 'sovereign' states can do on their own.

Why Study Politics?

So, the obvious answer to the question why study politics? is that it is intrinsically valuable because of its central place in the affairs of humans. More instrumentally, the skills ordinarily acquired by politics graduates (written and oral communication skills, some basic training in quantitative methods, problem solving, time-management and so on) are valuable for a whole host of careers. Particular suitable for politics graduates are careers in education, local authorities, the civil service and any other public service, although many also work in the private sector.

In addition, at Leicester, and in the UK higher education sector more generally, there has been a more recent emphasis on employability which includes not just a focus on the skills identified above, but also direct engagement with the workplace and specific employers. For instance, at Leicester we have recently introduced a module part taught by British parliamentary officials, and internship opportunities are now available for students to take advantage of either in the vacation or as part of a four-year degree programme.

Main Research Clusters

The Department of Politics and International Relations at Leicester has three main research clusters focusing on (i) intelligence, security and strategic studies, (ii) parties, participation and public opinion and (iii) global ethics and political theory. The research clusters aim to promote high quality research, act as a stimulating and supportive research environment for academic staff and PhD students, and facilitate collaborative research within the Department, and with other departments across the University. In addition, they promote and disseminate research through workshops, seminars, conferences and a website.

The Department has long had strengths in International Relations. The current focus is in the areas of intelligence, security and strategic studies. Members of the Department have played a central role in the development of Intelligence Studies as a distinct social science project. Other themes examined by members of the cluster are British foreign policy, nuclear strategy and non-proliferation issues, particularly challenges to the utility of nuclear weapons and nuclear deterrence, and critical security studies.

The Parties, Participation and Public Opinion (3PO) cluster brings together researchers working on themes related to political parties, political participation, public opinion and attitudes. Cutting across the core 3PO themes, members of the group have interests in Euroscepticism, migration, and parties in legislatures. A major five year European funded research project on governments' responsiveness to citizens' demands and preferences between elections - Democratic Responsiveness in Comparative Perspective (ResponsiveGov) - is run by members of the Department (http://www2.le.ac.uk/departments/politics/research/responsivegov).

The Global Ethics and Political Theory cluster coalesces around questions of injustice in general and, more specifically, the boundaries of belonging, and on the fundamental question of who counts. The work of the cluster considers how humans distinguish themselves from borderline cases including non-human animals, children and slaves.

My own work focuses on the normative value and political representation of nonhuman animals. My recent work on animals and justice is the first to consider the relationship between theories of justice and the protection of animals and the first that attempts to apply both ideal and non-ideal theory to animal ethics and, in particular, to develop a viable non-ideal theory of justice for animals (Garner, 2013).

Relevance of Academic Research in Politics and Available Scholarships

The Department welcomes applications from those with an interest in conducting doctoral research in any of these areas. From time to time we have scholarships available for graduate study. These are advertised on our web page:
http://www2.le.ac.uk/departments/politics/postgraduate/research-degrees).
An innovative feature of our provision is that we offer Master's programmes and research supervision through distance learning routes as well as the more conventional campus-based delivery. Details of distance learning can be found at:
http://www2.le.ac.uk/departments/politics/distance-learning.

All academic departments in the British higher education sector are now judged on the impact of their research. Clearly, academic research in politics can have a considerable societal and economic impact. It can, for instance, have a direct impact on public policy or it can have an indirect impact through altering the character of a debate within civil society. Interest groups and political parties are always on the look-out for research findings that they can utilise to further their cause.

To give one example, research on the ideological stance of UKIP (United Kingdom Independence Party) members and supporters, by academics within my own Department, has helped to shape the party's electoral strategy. More specifically, the research revealed that UKIP's support came not just from disillusioned Conservative voters but also from disillusioned Labour voters. UKIP's recent startling electoral success is at least partly down to their recognition (shaped by the research) that they should target Labour as well as Conservative-held parliamentary seats.

To give another example, my own research has been utilised, by the World Society for the Protection of Animals and by the British Department of the Environment, Food and Rural Affairs, to lobby the International Whaling Commission (IWC). Traditionally, this international organisation has regarded whales as merely an economic resource. Utilising my ethical typology, the British Government has sought to persuade the IWC to take more seriously the welfare of whales and not just their survival as a species. Central to this is the recognition that whales have an intrinsic value above and beyond their use to us as a resource.

References

Garner, R. 'Animal Welfare, Ethics and the Work of the International Whaling Commission',

Journal Of Global Ethics, 7, 3, 2011, pp. 279-90.

Garner, R (2013) A Theory of Justice for Animals Oxford: Oxford University Press

Lynch, P. and Whitaker, R. (2011) 'Explaining support for the UK Independence Party at the

2009 European Parliament elections', Journal of Elections, Public Opinion and Parties, 21:3, pp.359-79.

Chapter 13: Why Study Political Science? – Thomas Sutton, PhD

Thomas Sutton, PhD
Professor of Political Science
Baldwin Wallace University

Dr Thomas C. Sutton is a professor of Political Science, holds the Burton D. Morgan Chair in Entrepreneurial Studies, and is Director of the Community Research Institute at Baldwin Wallace University (BW) in Berea, Ohio. Professor Sutton teaches U.S. government and politics, public policy, and research methods. He has also developed an interest in West African politics and governance after spending a semester teaching and conducting research at the University of Cape Coast, Ghana with the support of a Fulbright fellowship. In this capacity, Professor Sutton conducts research and teaches courses in African politics and governance and African-American studies through his work with a newly established undergraduate minor in Africana Studies at BW. Professor Sutton's primary research work is through the Community Research Institute (CRI), which conducts surveys and organizational analysis for local government and nonprofit entities in Northeast Ohio. The mission of the CRI is to engage students in applied public policy research that benefits the communities of Northeast Ohio. The CRI has conducted projects for the Cuyahoga County Board of Elections, the Northeast Ohio Regional Sewer District, the County Public Library, and the Downtown Cleveland Alliance. Professor Sutton has published academic research in education policy and in studies of the executive branch during the Gilded Age, including the edited volume, ***Vice Presidents of the Gilded Age: A Documentary History*** *(2007: Mellen Press), and entries in U.S. Presidents and the Constitution (forthcoming in 2015: NYU Press). Professor Sutton also serves as an independent political analyst for ABC affiliate WEWS Newschannel Five. He lives with his wife Colleen and their daughters Kathleen (17) and Janelle (13) in Lakewood, Ohio.*

Political Science: A 'Questionable' Choice as a Field of Study?

Why do undergraduate students choose majors or minors in the field of political science? In an era of constant pressure to go to university, to prepare for a career in a field with good hiring prospects, to earn enough income to pay off student loans, buy a car, buy a house, and live comfortably in an expensive society, it is little wonder that political science can appear to be a 'questionable' choice as a field of study for undergraduate students.

As an undergraduate student, I studied politics in an interdisciplinary program in Political and Social Thought at the University of Virginia. After thirteen years in nonprofit management, I earned a Ph.D. in Political Science and Public Policy at Kent State University and began teaching at Baldwin Wallace University. I have often had to answer the question, 'what do you DO with a political science degree?'

As a liberal arts field, political science is challenged by this question, particularly when so few of our students choose to do the most directly related work of campaigning for elective office or securing a position in government. Students enjoy studying politics, from local issues to global affairs, but only a few are motivated to work on campaigns or run for office. Many would like to work in government, but this often requires earning a master's degree in public administration, public policy, or a law degree (juris doctor). My challenge as a professor and department chair has been to help students see the broader picture of opportunities in careers for which a political science degree is good preparation. It has also become evident that many careers require a master's degree, for which good preparation in undergraduate studies is important.

Why Political Science is a Good Field of Study?

The goal of this essay is to help students understand why political science is a good field of study as an endeavor in intellectual development, improvement of necessary career-related skills, and an enjoyable pursuit in and of itself. In short, I hope to offer a response to the perennial question I have faced over the past thirty-five years: 'why study political science?'

I teach in the Political Science Department at Baldwin Wallace University, an undergraduate institution in Berea, Ohio (on the western border of Cleveland) that serves 3,700 students, and includes Master's level study in the fields of Business Administration and Education. Our department has six full time and four part time faculty, with approximately 200 students majoring or minoring in the areas of political science, international studies, public service leadership, or legal studies. We also offer courses that are part of majors and minors in Criminal Justice Studies, Education, Leadership Studies, Sustainability, and Urban Studies.

Students choose to major in political science primarily because they are interested in the material. These are students who have an interest in the work of government, the dynamics of political campaigns, and the dramatic events of global affairs. Some are students with a heritage from other parts of the world who are trying to understand the countries where their families came from, in the context of being residents or citizens of the United States.

Traditionally, there is a cohort of students who choose political science as a field for preparation to attend law school (although this is by no means a requirement of law schools). Students choose political science because they are interested in social change, and want to learn how this is accomplished through politics and public policy.

In recent years, there has been a growing interest in this area among our students, with increasing numbers engaged in volunteer service in the nonprofit sector. These students want to engage in making their community and the world a better place. They study sociology, political science, philosophy, and related liberal arts fields, and often move into entry-level positions with nonprofit organizations, or study public policy or nonprofit management in a Master's program.

What do Students Learn to Do in Political Science?

The fundamental question remains: what do students learn to do in political science? What skills do they gain that make them attractive potential employees and candidates for graduate study? Based on the nature of the material studied and the typical assignments given in our courses, students in political science develop strong skills in writing, research, oral communication, and analysis. These are skills that every graduate program and employer seeks. The beauty of this skill set is that it will always be in demand. Technology helps to facilitate writing, presentations, and analysis, but it can never replace these fundamental human endeavors. A good undergraduate program in political science keeps abreast of technology developments and incorporates these into its coursework so that students will be prepared to use these tools.

Most importantly, students will be prepared to think in terms of change, developing a disposition to adapt to a continually changing set of circumstances in the workplace, in their communities, and at the national and international level. This mindset is a key component of long term career success. It is also a highly valued skill that can be used to help organizations to adapt to change.

The paradox of human existence is that while we live in an era of constant change, there are elements that persist, and need to be understood as the foundation stones of our lives and communities. In the Asanti culture in Ghana, there is a symbol of a bird looking backwards called the sankofa. The symbol means that in order to understand where you are going, you need to know where you came from. Political science includes the study of political theory, the work of political philosophers and theorists from the time of the Egyptians and Greeks through the present.

Political theory asks, 'what is human nature?'; 'what constitutes the good society?'; 'what is the meaning and practice of justice, equality, freedom?' These are perennial questions that offer frameworks of understanding that can be applied to contemporary challenges. Political science is also rooted in the study of history. One cannot understand current politics in any country, and especially between countries, without understanding the history of these dynamics. One cannot develop a full understanding of a country's politics without familiarity with its history, social structures, economy, and culture.

How to Conduct Empirical Research and Analysis

As one of the social sciences, political science requires an understanding of how to conduct empirical research and analysis. The field requires students to understand how to develop an objective stance when studying particular questions. Learning how to frame research questions, develop a study, collect and analyze data, and develop conclusions and recommendations are critical elements in the field (hence, the inclusion of 'science' in the field's name). Students at Baldwin Wallace University are required to take a course in research methods and analysis, in which they learn the rudiments of developing a research study, data collection, and analysis using quantitative analytical techniques. It is often a surprise to our students that they have to 'do math' in political science (e.g., use basic descriptive and inferential statistics).

I ask students how they expect to determine the outcome of an election, figure out an agency's budget, or grapple with the effect of the national debt, public spending, and taxation on the national economy without using math? What students should understand is that these skills are among the most highly valued by employers. A research study is like project management; figuring out the objectives, how these should be accomplished and developing and implementing an action plan are essential tasks in all organizations. Understanding how to analyze and use statistical data is a distinguishing skill that employers value. It is also a practical personal skill in a world where such data is used to encourage consumers to purchase a host of products in medical care, foods, and personal finance.

The work of students in classes is important, but must also include practical experiences such as internships, study abroad, and research presentations and publication. These are the types of experiences that help students gain a better understanding of their career opportunities and interests, and are also attractive to employers who identify specific activities as involving the use of particular skills. Baldwin Wallace University requires all of its graduates to engage in experiential learning.

Study Abroad and Internship Opportunities

Students in our political science programs study abroad and/or in Washington, D.C., are active in our Model United Nations, Model NATO, and Model African Union teams, volunteer with local nonprofit organizations, and conduct applied research for contract nonprofit and government organizations through our Community Research Institute (you can learn more by visiting our website: (http://www.bw.edu/academics/pol/). Our faculty collaborate as sponsors, mentors, and co-authors with students in undergraduate research papers that are presented at regional and national conferences, and have been published in undergraduate research journals. Our students have completed internships in local government, with nonprofit organizations, and with elected officials, include the Mayor of Cleveland, our Ohio Senators, members of Congress, and state elected officials. Our department also works closely with the BW Center for Innovation and Growth, which offers a variety of programs and applied research projects in the areas of innovation and entrepreneurship. These kinds of experiences are very important.

Students learn through experience what kinds of work are involved in different jobs and organizations and are able to better discern how they would like to spend their time in a career. Employers and graduate schools see the skills and talents of applicants in an applied setting when these kinds of experiences appear on a resume or graduate application.

Back to the original question: 'why study political science?' If you want to develop a foundational understanding of the world, desire to help make this a better place for others, are passionate about politics and issues, then political science is a good choice as an undergraduate major or minor. If you want to be valued as a potential employee of a nonprofit organization, government agency, or even a business, political science will help you to develop strong skills in writing, oral communication, research, and analysis. Rest assured, if you grasp the answer to 'why study political science?', the question of 'what do I do with a political science degree?' will be answered by your own interests, skill development and ability to seize opportunities as they arise.

Chapter 14: Political Science Is to Do Beyond Political Science – Professor Hyun Yun, PhD

Hyun Jung Yun, PhD
Associate Professor
Department of Political Science
Texas State University

Dr Hyun Jung Yun is an associate professor in the Department of Political Science at Texas State University. She earned two doctoral degrees in two different disciplines, one in Political Science and the other in Journalism and Communication from the University of Florida. Her research has dedicated to interdisciplinary approaches across political communication, public opinion, geopolitics and applied methodology, focusing on e-government, ethno-politics, social capital, and policy attitudes. Her publications in several leading journals, such as American Behavioral Scientist, Journalism Studies, and The American Review of Politics, and several book chapters demonstrate how individuals' political perceptions, attitudes, and behaviors are influenced by political predispositions within a group and by political resources within a given geopolitical and media system at the aggregate level. She has also received several grants from domestic and foreign governments, such as the U.S. Department of Education and the Korean National Assembly Research Service. As part of the grant-supported projects, her research teams developed: databases on U.S. election laws; educational modules for journalists to use as tools in order to be prepared to cover emergent news; and a certificate program in Southeast Asia Studies at Texas State University.

Beyond Direct Application

There has been frequent news coverage about the unpopularity of political science as a major when it comes to a lifetime career in some capitalist countries. In America, we have often heard that people undervalue political science as a career in that there has been more emphasis on the practices and research of hard science, which brings in direct revenues and tangible products, especially during economic downturns. However, it is a myth and it is misleading information. The reason why political science as a lifetime career is often underscored and devalued in capitalist countries, such as the U.S., is that people tend to simply calculate only monetary inputs and outputs by considering outweighed educational costs, and short-term income and unemployment under economic crises.

Let's think about the outputs and final objectives of majoring in political science in two broad perspectives: one is in the practical and direct outputs, and the other is in indirect applications to broader career fields. In a practical and direct perspective, if you want to work for the government, public agencies, international organizations, or non-profit organizations, studies in political science serve as the required qualification. As many of you may have seen in job advertising descriptions for those positions, political science is listed as a required degree, or at least as a relevant qualification and background.

However, a more important reason is that the study of political science gives you tremendous opportunities for diverse career paths. As you observe from everyday events, other fields of study are rooted in political science. For example, economic problems are often initiated by bad government decisions and inefficient policies. A nation's plans and directions for technology, future energy, and scientific research are all determined by the relevant government agencies and political leaders. In other words, the study of political science serves as the basis and foundation of other social and hard sciences and provides the reasoning for why and how those fields of study exist.

The study of political science has evolved and moved in the direction of both traditionalism and progressivism on the continuum of pragmatic and symbolic applications, depending on the dimension of issues and situations. For example, comparative politics, one of the subfields in political science, teaches students practical knowledge about different countries and regions, as well as symbolic values of diversity, mutuality, and co-existence. Whether you work within the field of political science, such as with an international organization, or in other fields, such as in the medical arena, you will need to handle situations to solve issues and problems by applying both practical knowledge and symbolic interpretation from multiple perspectives. The study of political science teaches you how to do politics and how to apply the knowledge of political science to diverse and transcending political contexts in the real-world career.

Finding Yourself and Understanding Others

Another important reason of majoring in political science is to be a qualified world citizen in the new political era. We now live in a wide-open, diverse, and intertwined world. We have observed that the barriers and territories of nations rapidly fade away; any political, economic, and social issues that countries face are related to other countries, and any decision one makes influences others. In the current world, there is no such thing that has an absolute value and meaning. The value of a thing or being can be determined in a relative term. "Who you are," "what that is," and "how it is" are understood by facts such as who evaluates you and things under certain conditions with certain perspectives.

The study of political science is to understand yourself, others, and the relationships between you and others. There are different and multiple cultural, economic, and social aspects of a fact or event. Research in political science provides you with the methods and tools to understand issues raised from inside and outside of your own culture and country, utilizing different perspectives and views of others. In human history, positive interaction and harmony, and negative conflicts and wars, have occurred when understanding or misunderstanding yourself and others. If you observe, evaluate, and judge others through your own values and perspectives, you are more likely to have conflicts with others. Without the skills of understanding and accepting differences, we end up with unnecessary conflicts and fights. That is the main reason for history's wars. If you understand others from their perspective and from being in their shoes, you are more likely to make friends. That is how different countries become political and economic allies and partners in harmony and peace.

The theories of political science explain who you and others are as citizens of certain countries, cultures, and histories at the macro-level of the group beyond the micro-level of the individual. For instance, consider the horrible situation of the Nigerian girls kidnapped by a terror group in 2014. As an American who is educated and who grew up in a democratic country with the values of liberty, freedom, equality, and humanity, the situation is hard to understand. The lack of the Nigerian government's actions and cooperation with other international support organizations is surprising. The relationships between the government and its people, the government and its military, and the government and other anti-political groups are obscure. The concepts of human rights and equality and the meanings of family, government, and tradition are very different from Western democratic countries. In addition, the infrastructure and economic power of Nigeria is unique in its path of development. Therefore, despite the violation of human rights and the lack of government responsibility, we are not supposed to judge or condemn anything from our own views.

The study of political science is to find yourself and understand others, and it teaches people diversities, differences, and tolerances so that we can help others with their needs. For instance, the subfields of political science – political theory, international relations, comparative politics, and political communication – explain the reasons for the political and economic underdevelopment, pre-existing interactions with others, unique and different social values, and slow and inefficient communication channels in Nigeria. Political science prepares you to be a better international citizen who can understand others from their perspectives and understand multiple interpretations of the real world. The world is getting smaller and closer, and this skill is a necessary quality to live in the current world.

Floating with Waves of Changes

Human history never stops somewhere in time. Knowledge of political science is concerned with "why and how things have happened and what will occur next," and then answers the questions with comprehensive and multiple aspects of diverse ideas. Let's think about the Cold War era – before, during, and after. What are the important factors that have driven each time period? In the pre – Cold War era, the military power that enabled countries to expand their territories was the core factor. During the Cold War era, ideology generated all political, socioeconomic, and cultural interactions and formed allies and enemies. Since the Cold War has ended, economic power has driven every part of human interaction and history.

The study of political science provides you with the logic and reasoning for why certain historic events occurred, how those events progressed and concluded, and where we are going from here. In other words, political science teaches you the past, current, and future trends and lets you float with the timely political, social, and economic waves. Without understanding the chronological changes in trends, you cannot appropriately work with others and cannot be on the same page with others in the current time.

As I listed above regarding a more tangible benefit, the study of political science educates you not only to be a well-trained political researcher and practitioner in the field of politics, but also to be a phenomenal individual who knows how to apply such knowledge to other fields of study and practice. In more meaningful but less tangible perspectives, the knowledge of political science enables you find your real self and understand others under the given political and historic circumstances, and lets you live with others, adjusting to the changes in human history as a valued world citizen.

Chapter 15: Why Study Political Science? Answers for Homo Politicus and Homo Civicus – Professor Jeb Barnes, PhD

Jeb Barnes, PhD
Department of Political Science
University of Southern California

Dr Jeb Barnes is an associate professor at the University of Southern California (USC), where he teaches courses on law, public policy and institutional analysis. He received his BA in history from Brown University and JD from the University of Chicago Law School. After clerking for the US Bankruptcy Court in Chicago and practicing as a litigator in Boston and San Francisco, Jeb returned to graduate school and earned his Ph.D. in political science from UC Berkeley. Jeb joined the faculty at USC in 2001 and, from 2003 to 2005, was a fellow in the Robert Wood Johnson Foundation's Scholars in Health Policy Research Program. His writing focuses on the role of rights, courts and litigation in the policy-making process and mixed-method research strategies. He has published numerous peer-reviewed articles and five books: ***How Policy Makes Politics: Law, Courts and Rights and the Struggle over Injury Compensation*** *(Oxford 2015) (with Thomas F. Burke),* ***Finding Pathways: Mixed-Method Research for Studying Causal Mechanisms*** *(Cambridge 2014) (with Nicholas Weller),* ***Dust-Up: Asbestos Litigation and the Failure of Common Sense Reform*** *(Georgetown 2011),* ***Making Policy, Making Law: An Interbranch Perspective*** *(Georgetown 2004) (co-edited with Mark C. Miller),* ***Overruled? Legislative Overrides, Pluralism, and Contemporary Court-Congress Relations*** *(Stanford 2004).*

What is Political Science?

Why study political science? This question is deceptively complex. To answer it, we must first define political science and then consider the question from the perspective of two very different types of people, what Robert Dahl once called "homo politicus"—individuals who are naturally interested in politics or self-proclaimed "political junkies"—and "homo civicus"—individuals who are responsible citizens but do not have any burning interest in the minutiae of politics or policy-making processes.[xxv]

Defining political science is a daunting task. It is an eclectic discipline at best, and a somewhat incoherent one at worst. It includes everything from philosophical inquiries into Plato to mathematical treatises on state-of-the-art statistical methods. Along the way, political scientists study nearly every aspect of political behavior, political institutions and policy-making processes in nearly every society in the world. Under these circumstances, we can only offer a working definition, recognizing that any definition will be problematic. Nevertheless, we need to start somewhere, so, for the sake of argument, let's assume that political science is the systematic study of the processes, principles, and structures of political institutions and government.

For homo politicus, the question of why study political science can be reframed as follows: why not just browse the web or tune into the media? In a digital age, we have unprecedented access to information about government and policy-making processes. Given these options, it seems a waste of time to read arcane studies by political scientists, especially when the peer review process ensures that scholarly work often takes years to see the light of day, making it seem hopelessly out of date in today's fast-paced world.

Who, *What*, *Where*, Why, and So *What?*

The answer stems from the systematic nature of political science. Media usually—although not always—offer a reasonable sense of the *who*, *what*, and *where* of current events, but these accounts are typically thin, providing episodic coverage that lacks critical context and rigorous research methods that give us greater purchase on the underlying causes of events and their broader implications. Put differently, media often fail to gives us any insight into the crucial questions of *why?* and *so what?*

For example, I recently wrote a book for undergraduates about the politics of asbestos litigation reform in the United States.[xxvi] Asbestos has been called a "magic mineral": [xxvii] it is a fiber made of rock, which is flexible enough to be woven into cloth yet is stronger than steel (as well as fireproof, resistant to corrosion, naturally abundant and easy to mine). It has been used in a dazzling array of products, including building materials, car parts, and consumer goods ranging from hair driers and children's modeling clay to soil conditioner. The problem is that exposure to asbestos can be lethal and literally millions have been (and continue to be) exposed to asbestos worldwide.

In contrast to other countries, the United States has relied heavily on litigation to address the problem of asbestos injury compensation.[xxviii] At first, litigation served heroic functions. It provided victims a means to bypass stingy government programs and receive some compensation. It also revealed decades of corporate wrongdoing and raised awareness about a major public health problem.

However, litigation has been a woefully inefficient and inconsistent means of compensation, which has failed to provide reliable relief to victims while dragging many companies that had little to do with creating the problem into bankruptcy. As the costs of litigation have continued to mount, judges have repeatedly begged Congress to provide a national solution to what has been called "the worst industrial accident in American history."[xxix] And yet Congress has failed—time and time again—to respond.

Homo politicus, especially those from the United States, probably would be vaguely aware of Congress' failure to act. Accounts of reform efforts were readily available and followed a familiar pattern. Advocates would introduce a bill, such as the Fairness in Asbestos Injury Reform Act (or FAIR Act), with some fanfare. But their efforts would eventually fizzle, usually because the bills would die in committee or proponents could not muster the needed two-thirds majority vote to overcome procedural requirements in the US Senate. Homo politicus would be tempted to file the story away as another instance of legislative gridlock in Washington.

This understanding would be incomplete at best and misleading at worst. It would be incomplete because the story of the fate of bills like the FAIR Act would give little insight as to *why* they failed. On its face, the FAIR Act promised to save victims and businesses billions of dollars in transaction costs by replacing litigation with a relatively streamlined compensation scheme and capping lawyers' fees.

Yet interest groups of all stripes badly divided over the bill. Nor would it explain why many prominent Democrats broke ranks with trial lawyers, their political allies, to push the bill, while conservative Republicans, political adversaries of trial lawyers, joined forces with them and fought it. Put simply, the fragmented politics of asbestos litigation reform defied conventional wisdom in ways that the media's sound bites glossed over. (The book argues that the key to understanding the puzzling politics of asbestos is seeing how litigation has shaped political interests over time, creating a patchwork of winners and losers in the courts who have cross cutting interests that divide them within and across political party lines.[xxx])

The conventional understanding would be misleading on the issue of *so what,* mischaracterizing the case's broader implications. Media accounts of the asbestos story tend to stress the theme of legislative gridlock, which implies policy inertia. Yet the case is a story of policy innovation not inertia, because courts have been proactive implementing what I have called "court-based tort reform" since the 1970s: the use of existing rules and procedures to change who decides, who pays, how much and to whom.[xxxi]

So, as congressional action has visibly ebbed, judicial policy-making has subtly flowed. From this vantage, the case of asbestos is not a simple example of legislative gridlock and the status-quo orientation of American politics; rather it is part of a much broader story of the judicialization of politics and policy-making in the United States and abroad. In sum, political science provides homo politicus something that is likely to be missing from the media: systematic accounts of the *why* and *so what* of politics.

The much tougher question is why should homo civicus study political science or at least take it seriously? It is one thing to ask political junkies to wade through systematic studies of the processes, principles, and structures of political institutions and government, but why should anyone else?

Two Reasons Why Homo Civicus Should Invest Some Time Studying Political Science

There are at least two reasons why homo civicus should invest some time studying political science. The first is that even if one is not keenly interested in politics generally, most of us care about some policy issues regardless of our ideological bent. If you are liberal, you are probably concerned about the effects of climate change. While this problem has many dimensions, including meteorological issues that fall outside the purview of political science, politics are at its center.

After all, many policy responses to climate change are well-known, such as imposing a carbon tax on activities that release greenhouse gasses, investing in alternative energy sources, and, especially in a country like the United States, consuming less energy. None of these measures is likely to "solve" the problem, but they could help. The main impediments to pursuing them are political not technical, and political science is uniquely poised to reveal the opportunities for (and constraints upon) pursuing policy change at the national and international level.

If you are conservative, then you are likely to be concerned about the size of government and the need for reforming welfare programs. Again, the problem is not technical. We know what to do: to bring government into fiscal balance, programs must be cut, taxes raised, and/or waste reduced. Deciding on the mix of these remedies is, of course, inherently political and, once again, political science can offer useful insights into the opportunity structure for reining in the costs of government as well as analyses of how different countries have tried to respond to this issue.

The point bears emphasis: to make a difference on issues that you care about, you must be able to identify policy options and read the institutional landscape to see when and where there are chances to act; political science offers a powerful lens for seeing the range of policy options and the institutional opportunities for action.

There is another, more general reason for homo civicus to consider political science, one that transcends any particular policy issue and speaks to the broader goals of higher education. We live in an age of enormous complexity and thus need strategies for making sense of modern life. For all of its faults, political science embraces diverse strategies for studying complexity because the processes, principles, and structures of political institutions and government are nothing if not complex.

On this score, political science's eclecticism is a disciplinary strength, not a weakness. By studying political science, homo civicus can gain exposure to diverse analytic methods that run the gamut from highly deductive modeling and lab experiments to careful ethnographies and historical case studies.

Studying these strategies in depth can give students tools for understanding the world around them and isn't that why we studying anything?

Chapter 16: Why Study Political Science? – Professor Greg Flynn, PhD and Ahmed Shafiqul Huque, PhD

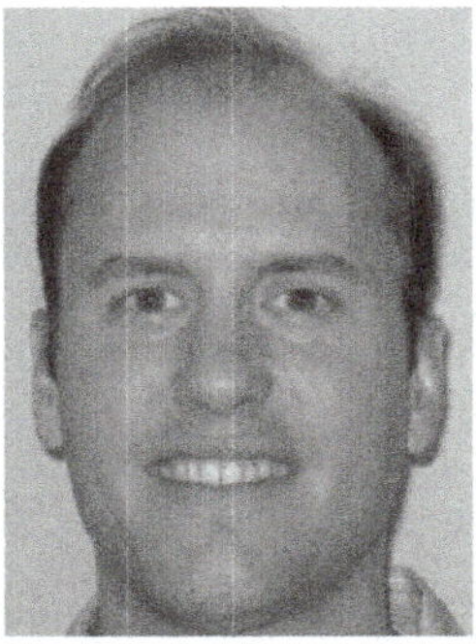

Greg Flynn, PhD
Assistant Professor
McMaster University

Dr Greg Flynn is an Assistant Professor at McMaster University. His teaching and research interests include Public Service Leadership, Public Law, Theoretical, Constitutional & Institutional Issues, and Issues in Canadian Politics.

Ahmed Shafiqul Huque, PhD
Professor and Chair of the Department of Political Science
McMaster University

Ahmed Shafiqul Huque is Professor and Chair of the Department of Political Science at McMaster University, Canada. His teaching and research interests are in the areas of public policy and administration, development, reform and governance. His recent publications include 'Public Sector Reforms in Developing Countries' (Routledge), Asian Immigrants in North America with HIV/AIDS (Springer), and Managing Development in a Globalized World (CRC Press).

Who Gets What, Where, When, How and Why?

Political science has been historically defined in a relatively narrow fashion as being about the study of resource allocation in society - who gets what, where, when, how and why. While vague and somewhat all encompassing, more recent answers to the question of "what is political science" have not done much better. Rather, they have provided an even more expansive approach, while at the same time being somewhat circular, defining political science as the study of politics, government or political phenomenon or the use of power in society. Both sets of definitions highlight the broad and diverse nature of the discipline, but neither is particularly insightful or helpful for a student trying to choose a field of study as they embark on a university career. However, after digging a little deeper, we can see that what falls under the study of political science as an academic discipline becomes not only much more meaningful in terms of what it incorporates, but also how critically important political science is to the world as a field of study.

Political science is not shy in seeking answers to big questions - questions that involve significant issues that challenge nations and communities as a whole and that present challenges to political actors and policy makers in the 21st century. Such important and ongoing research topics have included a number of highly salient national and international concerns that dominate our news headlines, such as: why do revolutions or overthrows in government occur (or perhaps more importantly, why they do not)? is democracy better or more effective than authoritarianism (and, if so, at what and why)? what can governments do to prevent or minimize the effects of national and international natural or man-made disasters? what impact do specific or general social, health, environmental, economic and/or trade policies have on communities, broader regional and national populations and economies? why citizens of some countries enjoy freedom and not others? why some countries remain poor while others prosper and, what identities and peoples constitute proper citizens in a community and how have those perceptions changed over time and with different governments? Research into these types of relevant and extremely contentious subjects go to providing a better understanding of the very foundations of our societies and call into question our reasons for joining together as political communities.

At the same time, political science also tries to critically explain individual events and individual behaviour that are also part of our broader political life. Issues such as, why do voters vote (or not) and what impacts the choices they make of who they vote for, which candidates are best suited for political appointment to the courts or other state agencies, do particular government policies promote or dissuade people from working or employers from hiring, all continue to remain prominent concerns in democratic countries and political science helps provide more localized understandings of specific political phenomenon and occurrences. It is a broad and diverse academic discipline that focuses on both macro and micro level themes and is concerned with the governance of people at all levels of organization and not solely with the actions and policy choices of large or national governments. Rather, it seeks answers through the examination of the formal organizations of the state and the informal organizations of society, the market, the family or the community. It is also interested in understanding policy and political impacts at all levels of organization – how local, regional, national and/or international political actors, acting alone, in concert or at cross-purposes, affect our lives on a daily basis.

In short, political science is about trying to critically understand the way the world operates. It helps us understand the role of government and national institutions and the role of citizens in the system There is an opportunity to better appreciate how communities are organized and governed, how power, force and violence (and different types of each) are used to control people and contest political outcomes, how government and governing institutions are created, transformed and circumvented and the roles they serve in maintaining political peace and stability, and how identities of people are formed, contested and reformed. It is made up of a number of broad sub-disciplines that cater to specific interests – international relations, comparative politics, political theory, public policy, public administration and country specific studies (Canadian, American, Indian, Russian or whatever country you are interested in).

Why a Research Degree in Political Science?

A research degree in political science offers an opportunity to explore problems and questions that are dynamic, important and relevant. As is apparent from defining political science in such broad terms, the discipline lends itself to a variety of research interests, both historic and current, and methodologies. Current and relevant research priorities in the discipline tend, in many ways, to mirror the local, national and global events and problems that shape our communities and that are at the forefront of the news. In this regard, the continuing integration of the international economy and its corresponding impacts (collectively falling under the general term "globalization"), the mass flows of migrants from country to country, region to region and continent to continent, and the increased securitization and militarization of domestic and foreign/international policies as a result of concerns associated with terrorism or nationalism highlight much of the emerging priorities in research in the fields of international relations and comparative politics. Domestically, studies on public policy, public administration and country specific fields have tended to focus on policy specific issues, with an increasing emphasis on policies related to social welfare, the environment, equity, gender and inclusion, as well as electoral and government reform in increasingly democratically challenged environments.

Reflecting the focus in these areas is an increasing trend of cutting across the traditional sub-disciplines and to incorporate a range of approaches, interests and methodologies into our understandings of politics. For example, the effects of globalization and other international concerns have spread into traditional domestic policy concerns resulting in emerging areas of study such global public policy, the internationalization of labour, civil rights, and human rights organizations and other broader social movements and the importance or irrelevance of domestic and international laws and governing organizations.

At McMaster, we are on the forefront of many of these emerging trends in research in political science. Our faculty research is primarily located in the sub-disciplines of international relations, public policy and public administration, with a particular focus on globalization and trans-nationalism, critical international relations, global political economy, social policy, multiculturalism and cultural policy, and labour rights, policy and issues. In addition to the Institute for Globalization and the Human Condition, McMaster has established a working group focusing on social policy in an age of austerity, a research cluster on labour rights and another on manufacturing economy, particularly in relation to the auto industry, and a specialization on global public policy.

In addition to our place as one of Canada's leading research departments, we also provide a range of funding opportunities for Canadian and international students, including general scholarships, grants and bursaries for students enrolled in the program as well as research specific grants designed to assist our students with the ability to undertake meaningful field research in the completion of the graduate and undergraduate degrees (www.graduate.mcmaster.ca/graduate-scholarships).

Student Life is also about Learning and Living

Student life is not only about what you courses you take and what you study, it is also about learning and living. As an academic discipline, political science is amenable to a wide range of teaching approaches that make it a unique, compelling and enjoyable academic discipline. While traditional lectures continue to make up a large component of class instruction, other pedagogical approaches are employed throughout the discipline, including the use of simulations, gaming and role play, problem based learning, and experiential and community based activities.

At McMaster, we have a number of innovative, award winning faculty members who employ such innovative pedagogical approaches and link their research with the classroom experience at both the undergraduate and graduate levels. These types of classroom activities provide students with the opportunity for practical, real life based learning experiences. We provide course specific knowledge of political events and outcomes that are current and relevant. But more than that, a degree in political science at McMaster can provide students with the skill sets that are in demand by employers around the world. While technical skills and specific knowledge are important, employers know that these abilities can be developed for specific jobs. Employers in both the public and private sector are looking for new employees who possess critical thinking, presentation and communication skills, creativity, writing, leadership, and teamwork skills.

Employment Opportunities for the Graduates

Coinciding with this broader and relevant skill development and the need to not just study, but to live, is the broader applicability of the study of politics beyond the completion of the academic degree. Not only is political science fascinating and highly relevant as an academic discipline, but it also presents a gateway into a range of interesting careers and opportunities in Canada and around the world. McMaster students regularly go on to further their academic foundations through attending law school, teachers college, business administration programs or advanced academic degrees. We have courses specifically designed to provide students with the skills, opportunities and experiences that lead into careers in government and the broader public sector.

Graduates from our undergraduate and graduate departments obtain employment in government at the municipal, provincial and national levels, market research firms, policy analysts for businesses and lobbying organizations. Their successes speak to the importance and relevance of political science, not just in understanding the world, but contributing to understanding and participating in it as well. While career paths may vary, a sound grounding in the discipline of political science contributes to the development of good citizens and results in benefits to the society.

Chapter 17: The Value of Political Science for Political Participation – Professor Zehra Arat, PhD

Zehra F. Kabasakal Arat, PhD
Professor of Political Science
University of Connecticut

Dr Zehra F. Kabasakal Arat is professor of political science at the University of Connecticut and Founding President of the Human Rights Section of the American Political Science Association. Her research focuses on human rights, with an emphasis on women's rights, and explores their relation to democracy, development, and globalization. She examines both normative and empirical questions that have worldwide implications, as well as the conceptualization of human rights in the Islamic context and their practice in the Republic of Turkey. Her publications include numerous journal articles and book chapters, as well as books, such as ***Democracy and Human Rights in Developing Countries**, **Deconstructing Images of "The Turkish Woman"**, **Human Rights Worldwide**, **Non-state Actors in the Human Rights Universe**, **Human Rights in Turkey**, and **The Uses and Misuses of Human Rights**.*

Politics: Struggle for Power

Political Science involves systematic study of politics. Students of "politics" have conceptualized the term differently, but a common broad understanding of it can be simply summarized as "struggle for power."

Power is usually defined as the *ability* to influence the behavior of others. Seen as conflict-oriented, this "power over" notion is rejected by critical and ethical theories of power. Redefining power not in terms of restrictions of others' behavior but as self-development, such theories introduce a counter and positive notion of power: one's *capability* to do things. This capability, or "power to," approach is linked to the discourse of empowerment (Squires 1999, Chapter 1). If power is defined in non-conflict terms, empowerment of a person would then mean not taking away someone else's power but the development of his/her own capabilities (Nussbaum 2011; Hartsock 1996). Articulating power in development/capabilities terms, instead of domination/conflict terms, empowerment discourse rejects the notion that power is a zero-sum game.

Ability to Influence Others' Behavior and *Capability* to Choose and do What one Desires

However, "the *ability* to influence others' behavior" can be seen as the other side of "the *capability* to choose and do what one desires." Treating these as two sides of the same coin would bring us back to treating empowerment as a power struggle and power as zero-sum, which means the empowerment of the currently marginalized has to come at the expense of those who are already powerful. In other words, if one acquires the capability of asserting control over her own body and life – i.e., she is empowered – then those who used to have power over her would have less of it. [16]

Sources of power are many, including economic, political, physical (both muscles and weapons), institutional, traditional/customary (e.g., motherhood, seniority, age), psychological, collective/organizational, knowledge/experience, formal education, physical/erotic attractiveness, etc. While some sources may be finite, or near finite (e.g., money, seats in the parliament), others may be acquired by one without necessarily subtracting it from others (e.g., knowledge). Sources of power may not only overlap but also reinforce each

---

[16] However considering power as a zero-sum game does not reject the "positive" aspect of power, either. "Influencing others' behavior" would not necessarily mean making one do what she does not want to do, as it is commonly understood in the "power over" approach. Moreover, an important aspect of "exercising power" is engendering norms, values and institutions that determine the choices and domain of decision-making (Bachrach and Baratz 1962; Lukes 1978) and limit people's options as well as demands (Cohen and Rogers 1983, Chapter 3). On the other hand, one does not always exercise power to maximize his/her self-interest but also to make another person to take steps toward improving his/her capabilities. While we are familiar with this latter possibility in the context of relationships that involve more or less equal people (e.g., friendships) or caring ones (e.g., parent-child relationships), we should be able to imagine it in less intimate, societal contexts, as attempted by Kathleen Jones who introduced the concept of "compassionate authority" (Jones 1993).

other, and people have access to them in various degrees. In other words, power is everywhere, and no one is omnipotent or completely impotent.

Now, let's return to the definition of politics as "struggle for power." I would like to start with two important arguments related to this particular definition of politics. First, struggle for power takes place at *different levels* and at *different sites*. Second, *politics* are not about a set of events, such as elections, parliamentary debates or legislation, but should be understood as an ongoing complex *process* that involves multiple events and processes. In other words, struggle for power is a continuous process, as well.

Philosophers have *Interpreted* the World in Various Ways, the Point is to *Change* it

A famous quotation from a famous German philosopher would be helpful in elaborating on these arguments. In one of his early manuscripts, "Theses on Feuerbach," Karl Marx writes: "Philosophers have hitherto only *interpreted* the world in various ways, the point is to *change* it" (emphasis original, Marx 1845, paragraph 11). If we attempt to *change* the world – or simply our lives – we would need to be engaged in politics *consciously*.

Our engagement in politics is not limited to high politics. By "high politics," I mean the conventional understanding of politics, which refer to the activities that are related to governments: how governments are formed, how they reach certain decisions, how they implement those decisions and enforce law, how people influence government behavior, as well how governments interact with each other. The latter one is referred to as International Relations and is either studied as a sub-field of Political Science or treated as a different, though related discipline.

Ordinary people enter these "political" processes usually as the target of decisions. Laws, after all, are enacted to regulate our behavior. However, people also participate in the formation and operation of governments.

In addition to this government-oriented "high" politics, there is also what I would like to call *everyday politics*, which is subtler and less visible but have profound impacts on our lives. Everyday politics operate at our homes, at schools, at work . . . . . . We tend to study this type of politics under the titles of "gender politics," "race politics," "labor politics," etc. These relatively new interests of Political Scientists have been traditionally studied by Sociologists,

Anthropologists, or those engaged in interdisciplinary fields such as Cultural Studies or Women's/Gender/Feminist Studies.

At all of these sites – homes, schools, factories, office etc. – individuals or groups use their power sources, including certain rules and regulations that are institutional sources of power, to influence our behavior. We do the same and try to influence others' behaviors. In other words, we are all engaged in power struggles.

Moreover, these two types of politics – high politics and everyday politics – are intertwined. They always interact with each other, redistributing power sources and redefining power relations. Consequently, if we want it or not, we participate in both types of politics and live in a political world. Then the critical question is: "Do we want to be engaged in politics incidentally or consciously, by being aware of the consequences of our action and non-action?"

It is clear that if we want change, we need to be cognizant and take action. No action *may* mean maintaining the status quo. If one is content with her position in the society, the state of public affairs or the social conditions, she may choose not to take action. Yet, taking no action would not guarantee the continuation of the status quo. If she does not struggle to guard it, those who work against it may change the conditions in the direction of their preferences. This is why I refer to politics, or the struggle for power, as a continuous process.

Drastic Consequences of not being Conscious of our Action

Of course, those with more sources of power, whom we consider "the privileged ones," would be able to sustain the conditions that are favorable to them more effectively than those who have limited sources. Therefore, those with limited sources have to be engaged in politics more consciously and actively, and be more vigilant even to preserve what they have.

To highlight the drastic consequences of not being conscious of our action, or lack of action, I would like to conclude by citing another German philosopher, Hannah Arendt. Arendt uses the term "banality of evil" to describe what happened in Nazi Germany (Arendt 1963/2006). She notes that ordinary Germans, like Adolf Eichmann who was in charge of the transportation of Jews to concentration camps, participated in the Holocaust not because they hated

Jews or they were staunch anti-Semites. They simply continued with carrying out their jobs and living their lives without any reflection on the consequences of their actions or inactions.

Political Science offers neither a single moral guidance nor a road map to power. However, by introducing different political philosophies, case studies, common patterns, and analytical tools, it shows students that politics are everywhere and help them participate in the process more consciously and effectively.

References

Arendt, Hannah. 1963/2006. Eichmann in Jerusalem: A Report on the Banality of Evil. New York: Penguin Classics.

Bachrach, Peter, and Baratz. Morton S. 1962. "Two Faces of Power." Americun Political Science Review 56:4 (December): 947-952.

Cohen, Joshua, and Rogers, Joel. 1983. On Democracy: Toward a Transformation of American Society. Middlesex, England: Penguin Books.

Hartsock, Nancy. 1996. "Community/Sexuality/Gender: Rethinking Power," in N. Hirschmann and C. Di Stefano, eds. Revisioning the Political: Feminist Reconstructions of Traditional Concepts in Western Political Theory. Oxford: Westview Preess: 27-50.

Jones, Kathleen B. 1993. Compassionate Authority: Democracy and Representation of Women. London: Routledge Press.

Lukes, Steven. 1978. Power: A Radical View. London: Macmillan Press,.

Marx, Karl. 1845. "Theses on Feuerbach." Brussels, Spring1845. Marxist Internet Archive.
http://www.marxists.org/archive/marx/works/1845/theses/. (Accessed July 11, 2014).

Nussbaum, Martha C. 2011. Creating Capabilities: The Human Development Approach. Cambridge, MA: Belknap Press of Harvard University Press.

Squires, Judith. 1999. Gender in Political Theory. Cambridge; Polity Press.

Chapter 18: Why Study the Politics of Human Rights? – Professor Clair Apodaca, PhD

Clair Apodaca, PhD
Associate Professor, Department of Political Science
Virginia Tech

Dr Clair Apodaca, (PhD, Purdue University, 1996) is an Associate Professor in the Department of Political Science at Virginia Tech. Professor Apodaca's academic education and research include a broad background in international relations with a focus on human rights and human security. Dr. Apodaca's research attempts to understand the many multifaceted and interrelated causes of human rights violations and how those violations threaten human wellbeing, the nation-state, and international peace. Dr. Apodaca is the author of two books, ***Child Hunger and Human Rights: International Governance*** *(Routledge 2010) and* ***Understanding U.S. Human Rights Policy: A Paradoxical Legacy*** *(Routledge 2006) along with over a dozen articles published in some of the most prestigious scholarly journals on the complex issue of global human rights. Her work has appeared in the Journal of Human Rights, International Studies Quarterly, and Human Rights Quarterly, among many others.*

Subfields of Political Science

Why study the politics of human rights? Students who study the politics of human rights are interested in the nature of contemporary human rights problems and the role that politics and political regimes play in generating them, the economic, social, and political factors associated with human rights progress, and the strategies currently being employed to improve human rights in different settings.

The contribution by political science to the study and understanding of human rights is significant, both in substantive and methodological terms.

Human rights can be studied from each of the subfields of political science. International relations focuses on global issues, that is to say, worldwide political, social, or economic processes and inter-state dynamics, the mass media, or military intervention, global terrorism and counterterrorism, and their effects on human rights. In addition, international relations is also concerned with how foreign policy influences human rights policy in foreign countries.

International relations scholars would ask questions on how war, global trade and investment, development, or international organizations influence or impact human rights attainment. International or global investigations of human rights typically involve the use of quantitative methods and large data sets (what is known as large-n studies). These methods allows human rights to be studied over time and space, allowing the researcher to make inferences about the relationship among variables.

Comparative political science, on the other hand, is traditionally concerned with the determinants and outcomes of different political regimes and political cultures. Comparative politics looks at the similarities and differences across countries to develop a set of generalizations about human rights. Thus, human rights can be studied from a comparative perspective, examining how different political systems, cultures, and regimes interpret, promote and protect human rights. The comparative method is based on the cross-cultural comparison of a few countries using qualitative analysis or employing small-n quantitative analysis.

Those studying human rights from an domestic politics would investigate issues of civil rights and fundamental freedoms looking at how different government policies and programs impact the political and economic rights of domestic residents. Finally, political theory would allow students to contemplate the philosophical underpinnings and contemporary theoretical perspectives of human rights.

What is the Relationship between the Modern Nation State and Civil Society?

Through the study of political science, the student will gain a deeper understanding of the relationship between the modern nation state and civil society. Presently, over 3 billion people live in countries where their political, civil, economic and social rights are not adequately protected. Domestic and international human rights instruments provide the legal tools necessary to empower civil society in order to hold governments accountable for implementing human rights. Students who study the politics of human rights want answers for common questions such as: Why do some states commit more human rights violations than others? What role does political order and regimes play in state repression? How do power inequalities, discrimination, and poverty shape a state's commitment to combat human rights violations? How can one evaluate the implementation of human rights obligations in practice?

The development of human rights first achieved a prominent place in political science in the aftermath of World War II with the creation of the United Nations. While the Charter of the United Nations laid the conceptual framework for the development of a legally binding human rights law, it failed to clarify the substance of human rights. The Universal Declaration of Human Rights (1948) is a non-binding proclamation that represents the international acceptance of proper standards of conduct. The Universal Declaration was intended to clarify "a common standard of achievement for all peoples and all nations." The Universal Declaration provides an extensive catalog of rights all people possess as members of the human family.

For the first time, in 1948, human rights and fundamental freedoms were given explicit meaning. In order to give the principles articulated in the Universal Declaration of Human Rights greater clarity, the United Nations drafted a Covenant on Human Rights. The ensuing debates as to whether economic and social rights are of the same character as civil and political rights resulted in two separate and overlapping documents: The International Covenant on Civil and Political Rights, and the International Covenant on Economic, Social, and Cultural Rights. From these founding documents, there are now dozens of international human rights legal instruments.

State sovereignty is a core principle in politics and one of the most significant impediments to securing human rights. State sovereignty means that the state has the right to govern itself, has exclusive authority over its population and territory, the sole prerogative to exercise power and violence within its territory, and is free from external interference in its internal affairs. Thus, how a state treats its citizens is not a matter of concern for other states or the international community.

Yet, the concept of human rights challenges the sacrosanct precept of state sovereignty. Human rights are a direct challenge to the authority and sovereignty of states, because human rights limit the states' prerogatives to treat their citizens and residents as they see fit. Human rights trump traditional conceptions of state sovereignty since it allows citizens to limit the state's power and requires the state to protect and provide for its citizens.

Who Gets What, When and How?

Human rights apply to all human beings independent of the whim and interest of the state. Universality implies that human rights belong to all human beings regardless of where they live thus transcending state borders. States do not determine what rights individuals residing within their territory hold.

Furthermore, states often claim the priority of sovereignty over the universality of human rights and chafe at the implementation of domestic and international monitoring mechanisms needed to oversee human rights performances by states. Equality means that each and every human being is equally entitled to human rights and equal treatment in the protection and provision of such rights. Everyone has an equal entitlement to enjoy his or her rights without any unmerited discrimination. Human rights, based on the twin concepts of universality and equality, fit neatly in the well-known definition of politics as "who gets what, when and how" (Laswell 1950).

At Virginia Tech we have several course related to human rights, for example, women's rights, global human rights issues, human rights issues and perspectives, global poverty and welfare, Indigenous peoples and human rights, and constitutional law: civil and political rights. In addition, we have a human rights center that serves as a hub for human rights conferences and research at Virginia Tech. Students with a focus on human rights have found employment in the field of human rights with international organizations, such as the United Nations, non-governmental organizations, for example Amnesty International, and with the United States government particularly in the US State Department.

Human rights are an important issue in contemporary politics and will remain so for many years to come. Human rights are not neutral but inherently political in their foundation, creation, development, and application. Therefore, there are still many ideological battles and contentious issues related to human rights that have not yet been settled. Human rights are an important force in global, regional, and domestic politics.

Reference

Harold D. Lasswell, Politics: Who Gets What When and How (New York: P. Smith, 1950)

Chapter 19: Students should Choose a Major in Political Science for Several Reasons

David A. Dulio, PhD
Professor and Chair of the Political Science Department
Oakland University

Dr David A. Dulio is Professor and Chair of the Political Science Department at Oakland University where he teaches courses on campaigns and elections, Congress, political parties, interest groups, and other areas of American politics generally. Dulio has published nine books, including ***Campaigns from the Ground Up*** *(Paradigm Publishers 2015),* ***Cases in Congressional Campaigns: Riding the Wave*** *(Routledge 2011), and* ***For Better or Worse? How Professional Political Consultants are Changing Elections in the United States*** *(SUNY Press 2004). He has written dozens articles and book chapters on subjects ranging from the role of professional consultants in US elections to campaign finance. Dulio is also a former American Political Science Congressional Fellow on Capitol Hill where he worked in the U.S. House of Representatives Republican Conference for former US Rep. J.C. Watts, Jr. (R-OK).*

Political Science is a Broad Discipline

1. So, our curious students would like to first know what actually the academic discipline/major of Political Science is. How do you find the scope/future of Political Science in terms of career opportunities and options? Why should the students choose Political Science as their undergraduate or postgraduate major?

Political Science is a broad discipline that includes many sub-fields. While there are too many to mention here, a partial list could include international relations, comparative politics, regional or area studies – e.g., U.S. politics, Latin American politics, European politics, Asian politics, African politics, etc. – political theory, and many others. For someone who studies U.S. politics, like I do, there are also many fields within the one field; scholars study topics such as Congress, the presidency, campaigns and elections, political behavior, media and politics (including the Internet), public administration, public policy, political parties, interest groups, as well as others. The same kind of variety can be seen in other areas as well.

For instance in the field of international relations, scholars study topics like international political economy, foreign policy, international organizations, human rights, conflict resolution, and others. In the ever-changing global world in which we live, students with an understanding of these topics will be well-served.

I believe students should choose a major in political science for several reasons. First, as I mentioned earlier, the content you will study in any area of political science will be important for you as you engage as a member of our global society.

Moreover, and possibly more importantly, the skills that you will learn in political science courses are skills that can be used in just about any career or occupation. Many students who study political science look to go into careers in government sectors, law, "practical" or "applied" politics like campaigns and lobbying, education, non-profits, foreign service, the diplomatic corps, and other areas.

But the great thing about a political science degree is that you will also learn important skills like how to conduct research, how to become a strong writer, critical thinking, data analysis, and others. These are skills that employers in the U.S. report that they are looking for in new employees, regardless of their major in college. At its heart, political science is a liberal arts degree.

You can be a Part of the World of Making Great Discoveries about Questions and Phenomena that Interest you

2. Now, why should our prospective students undertake a research degree in Political Science? Can you briefly discuss the research areas or priorities in Political Science that the Oakland University is actively driving/pursuing please? What are your own research interests (and any interesting research findings)? Are there any scholarships available for our domestic and international students at the Oakland University [note: as relevant, please provide links to scholarship information/sources]?

One of the great things about studying political science is that there are so many questions yet to be asked or addressed. While scholars have certainly written a lot on many subjects within the discipline, there are always new and emerging areas that deserve study in addition to age-old questions that can be re-examined with new data or with a new or different focus.

For instance, in my area of study, campaigns and elections, scholars have studied questions centered on voting behavior for decades. But with each new election cycle, there is more and new data that can be analyzed and studied.

In addition, scholars find new ways to study questions that have been around for a long time. The emergence of the Internet, for example, has meant that questions about campaign communication, voter learning, candidates' strategic campaign decision making, and other aspects of campaigning are being revisited with a litany of interest results being uncovered. When you undertake a research degree in political science, you can be a part of the world of making great discoveries about questions and phenomena that interest you.

While we don't have a Ph.D. program in any of our three degree programs at Oakland University, our students do have opportunities to conduct their own research programs. First, in our Master of Public Administration program, students may elect to complete an independent research project as their capstone experience.

In our case, students have an opportunity to apply what they have learned and make a contribution toward solving an administrative or management-related problem; usually this is with a local government organization, a non-profit organization, or an organization a public administration focus. Our students design their research project with the help of a faculty member who has expertise in the area they are interested in studying and then conduct the research and write up the results themselves.

In addition, there is a growing culture of undergraduate research in the Political Science Department at Oakland University. In the last several years, more of our undergraduate students have taken the opportunity available to them to engage in research with faculty. Some work with faculty as co-authors while others work on independent research projects with guidance from faculty. In recent years dozens of our students have gone to professional political science conferences – such as the Midwest Political Science Association Annual Meeting and the International Studies Association Annual Meeting – accompanied by their faculty mentors.

Usefulness of Political Science Research

3. It has been said that the practitioners (non-academics/industry professionals) generally are less interested in taking advantage of academic research in Political Science. How does academic research in Political Science benefit them? How can Political Science research be made more relevant, useful and interesting to practitioners?

I do think there is some skepticism among political professionals about political science research. In some cases I think this is due to a lack of familiarity with the research that is conducted. In other cases I think this is due to the fact that some political professionals do not believe political scientists work in the "real world" and that academic research cannot translate to the areas in which they work. Interestingly, some practitioners – like those who conduct survey research or polls – are often trained in political science and have advanced degrees.

The research scholars conduct can be of great benefit to political practitioners. The findings from research projects can inform the work they do and even help them do their work. For instance, in the area of campaigns and elections, studies on different information sources (e.g., television, mail, Web-based sources) that are available to candidates to spread their campaign message and their effectiveness in conveying information to voters that they remember and internalize can help campaigners decide the best way to communicate with potential voters. In addition, the kinds of communication that are most effective can be studied and inform as well as help campaigners.

For instance, are "positive" television advertisements or "negative" advertisements more effective during a campaign? Scholars' research can help inform candidates' decisions and even the media's coverage of the campaign.

There are many more examples of this kind of usefulness of political science research throughout the discipline and all its subfields.

Chapter 20: Why I Love Studying International Relations? – Professor Kenneth Schultz, PhD

Kenneth A. Schultz, PhD
Professor of Political Science
Stanford University

Kenneth A. Schultz is professor of political science at Stanford University. His research examines international conflict and conflict resolution, with a particular focus on the domestic political influences on foreign policy choices. His most recent work deals with the origins and resolution of territorial conflicts between states. He is the author of ***Democracy and Coercive Diplomacy*** *(Cambridge University Press, 2001),* ***World Politics: Interests, Interactions, and Institutions*** *(with David Lake and Jeffry Frieden, Norton, 2013), as well as numerous articles in peer-reviewed scholarly journals. He was the recipient the 2003 Karl Deutsch Award, given by the International Studies Association, and a Dean's Award for Excellence in Teaching, given by the Stanford University School of Humanities and Sciences.*

Interesting Questions and Answers

When I was a young boy, a family friend promised to take my brother and me on a fishing expedition for tuna. I was excited, but my brother, who is a few years older and wiser, was skeptical. Despite the impression you might get from the grocery store, tuna don't fit in little cans; they are big, strong, fast fish and they fight long and hard before allowing themselves to be caught. My brother did a little research, and he concluded there was no way we were really going to catch tuna. Our fishing rods weren't sturdy enough, our fishing line wasn't strong enough, and the boat we were going out on wasn't big enough to bring in a tuna. When we confronted our friend about this, he admitted that, no, we weren't going to catch any tuna. In fact, given how fished out the waters of the bay were, it was quite possible that we weren't going to catch anything at all. But as long as we were going to come back empty-handed, it would be good to think that the fish we failed to catch were big ones.

That is why I love studying international relations. Sometimes we have good answers, sometimes we don't. But even when we don't, the questions we grapple with are big ones: Why are there wars? Why are some people and nations poor while others are rich? Why do some live in freedom, while others live in tyranny? Why do people sometimes take up arms against their governments or foreign states? Why is it so hard for people and countries to cooperate to stop genocides, resolve humanitarian crises, or prevent catastrophic environmental damage? Even though we won't end wars or poverty or misery, progress is possible, and incremental improvements in our understanding of these problems are a necessary condition for such progress.

A Policy Practitioner Needs an Answer Now

It is often said that those engaged in policymaking have little use for the research that scholars like me produce. It is certainly true that much of the academic literature is hard for non-specialists to read, sometimes very abstract, and/or focused on historical events that have little clear relevance to today's problems. A practitioner will not be able to open up the latest issue of top-rated academic journals like *International Organization*, *International Security*, or *International Studies Quarterly* and find the key to resolving the Israeli-Palestinian conflict or persuading North Korea to dismantle its nuclear program. Most scholars in this area have no direct line to policymakers to give them advice on such issues. We also operate on different time scales. A policy practitioner needs an answer now; a scholar likes nothing more than a problem that will occupy her for years.

So why bother studying this stuff? When asked this question, I often compare our relationship to policymakers to the relationship between physicists and engineers. The understanding of physics is crucial to building a bridge or skyscraper that won't fall down. And, yet, most engineers engaged in designing bridges and skyscrapers would find little guidance in what most physicists do.

An engineer needs to design a bridge now; a physicist is engaged in a long-run effort to understand the basic nature of matter and motion. Some of what they learn may have applications in the near term, while other discoveries may not have practical applications for some time, if ever. The insights from basic research trickle slowly and unevenly into the practical sphere. The interface between the two realms is porous: neither wholly open, nor fully blocked.

The most Lasting Impact of my Career

And one of the main mechanisms for ideas to flow across the divide is you, our students. I am fortunate to be a relatively accomplished scholar, with a good number of publications under my name. But when I think about what is likely to be the most lasting impact of my career, there is little question in my mind that it is through the students that I have taught. It is by observing the world, analyzing events, and then passing the tools we have developed and insights we have gained onto a next generation that our ideas seep into the public realm.

Why Game Theory is Important?

Let me give you an example of a tool that you could pick up that will help you as citizens, scholars, or future policymakers. Some of my research makes use of game theory, a set of tools used widely in economics and political science. Game theory provides a way to think systematically about strategic situations, in which the outcome depends on the choices of multiple actors with different goals.

When you walk into a restaurant, you face a choice: what do you want to eat? You consider which option on the menu you'd like the most, and you order it. It doesn't make sense to order something other than your most preferred option. But say you get to a four-way stop at the same time as another car. Your most preferred outcome is to get through the intersection first, but that does not mean that your best strategy is to go first.

The outcome of this interaction—whether you get through first or second or end up crashing—depends on what the other driver does. If you both try to go first, you end up with the worst outcome. In this case, if you think that the other driver will go first, your best response is to wait: after all, a small delay is better than a crash.

The more general point is that in strategic interactions, a player (it is *game* theory, after all) is almost always better off if she thinks not only about what she wants but also what other players want and, therefore, how they are likely to respond. Failure to correctly anticipate the actions of others, or to interpret the signals they try to send you, can lead to bad outcomes. Academic articles that rely on game theory are often indecipherable to those unfamiliar with the jargon or intimidated by the math. But I urge you to check some out, because the insights into strategic interaction are invaluable for anyone thinking about international relations or diplomacy.

Think Strategically

I mention this because, in my view, one of the biggest foreign policy mistakes made by the United States—the invasion of Iraq—resulted from a failure to think strategically. People in the Bush administration believed that the United States had the power and the will to topple the regime of Saddam Hussein and install a friendly government. Their faith in American power and the (supposed) attractiveness of American values and institutions led them to assume that groups within Iraq would welcome us and work with us and that other actors, like Iran next door, would be too cowed to meddle.

Sure, others might be waiting at that four way stop, but we had the biggest, baddest car, so why bother to anticipate what they would do? We would lead; they would have no choice but to follow. In making these assumptions, the administration failed to take seriously the interests of other actors or anticipate how they might respond to our invasion.

Instead of assuming that weaker actors would be cowed or defeated, they should have considered the strategies available to the weak—e.g., terrorism and insurgency—that can frustrate a stronger military power. A more thorough analysis of the goals and motivations of different groups within Iraq, and how little we knew about those groups, might have led to greater caution both in the decision to go to war and in the planning for the post-invasion environment. Ideology and overconfidence was a poor substitute for strategic thought.

Analyze What Went Wrong - Don't Make the Same Mistakes

So I was one of more than 850 international relations specialists who signed an open letter in October 2004 criticizing the war as "misguided" and calling for a change in course. It had no observable effect. So, again, why bother? Because, as students of international politics, we learn from these kinds of events and develop tools to understand them. We can analyze what went wrong, use the problems to better understand the challenges of dealing with hostile regimes or providing order in divided societies, teach our students to do the same, and help make sure that the next generation—hopefully, some of you—doesn't make the same mistakes. The process by which our insights trickle into the public sphere can be slow, long, and arduous, but ultimately worth it. Kind of like reeling in a tuna (or so I'm told).

Chapter 21: What's Studying Philosophy Good For? – Professor Hans-Herbert Kögler, PhD

Hans-Herbert Kögler, PhD
Professor of Philosophy and Department Chair
University of North Florida

Dr Hans-Herbert Kögler is a Professor of Philosophy and Department Chair at the University of North Florida, Jacksonville. His philosophical education happened in Germany as well as in the US, including Frankfurt University (Ph.D.), Northwestern, the New School, and UC Berkeley, working with professors as distinguished as Jürgen Habermas (MA and PhD advisor), Axel Honneth, Thomas McCarthy, Richard Bernstein, and Hubert Dreyfus. His philosophical work is in social and political philosophy, philosophy of language, and philosophy of social science, including hermeneutics and critical theory fused to a critical hermeneutics. He is the recipient of the UNF Outstanding Teaching Award, NEH fellowships at the University of Arizona and Boston University, a frequent guest professor at the Alpe-Adria University in Klagenfurt, Austria and Visiting Scholar at Charles University, Prague. Major publications include ***The Power of Dialogue*** *(MIT Press 1999);* ***Michel Foucault*** *(2004); Kultura, kritika, dialog (2006, the co-edited* ***Empathy and Agency: The Problem of Understanding in the Human Sciences*** *(2000); his influential and widely cited work on dialogue and social agency is further developed in over 70 articles and book chapters, written in English and German and translated into Czech, French, Russian, Italian. Kögler lectures on critical interpretation, cosmopolitanism, and intercultural dialogue via worldwide invitations, among them the University of London, Northwestern, Berlin, Vienna, Prague, Frankfurt, Auckland, and Pécs in Hungary. Recent contributions include "Empathy, Dialogue, Critique: How should we understand (inter) cultural violence?" in The Agon of Interpretations: Towards a Critical Intercultural Hermeneutics (2014) and*

"Ethics and Community" in the Routledge Companion to Hermeneutics (2015).

Why should you Embark on the Study of Philosophy?

When the question 'Why study philosophy?' is posed, the issue for many is the usefulness of philosophy for life. Usefulness, however, is often considered in terms of the subsequent employment opportunities that the academic study of philosophy may open up. The general perception is that philosophy rarely opens up such opportunities. Besides the reproduction of professional philosophers, which is to say professors of philosophy in higher learning, no such opportunities are considered to exist.

Philosophy as an academic study is seen as a luxury and an idle enterprise, if not a waste of time—in other words, it is considered useless. Now one can point out that those with academic degrees in philosophy fare quite well, and certainly do so in comparison to their peers in humanities and social sciences in terms of average salary, professional status, and employment satisfaction. Indeed, we will see that a rich and complex conception of the usefulness of philosophy will actually help explain why professionally trained philosophers usually land on their feet—and usually do so on pretty high grounds.

Yet this answer, while comforting to the students of philosophy, would already have accepted the narrow terms of the question. The issue of employment prospects immediately limits the 'why of philosophy' in at least in two major ways. The first one is rather obvious, as it entails an instrumental vision of philosophy bypassing the internal merit of the cognitive qualities of the academic study of philosophy. The second one relates to the focus on an individual person's life, which narrows the view by ignoring wider cultural and social benefits. In what follows I will show that philosophy carries at least two further functions, one with regard to culture, and the other with regard to society.

The 'dialectical' point of this essay is that overcoming the narrow focus on employment and earnings prospects actually allows us to widen the view to understand why philosophy majors in the end do get great jobs with good salaries. The knowledge, skills, and reflexive capabilities that are learned in the study of philosophy, while mostly not directly correlated with existing job descriptions, do equip agents to succeed in positions demanding higher-order thinking and communication skills.

And yet, besides the existentially relevant focus on job prospects, the bigger issue is that of the good of philosophy as such. My claim is that the beneficial nature of philosophy can only be fully understood if we see how philosophical thought and practice relate to cultural traditions as well as social interaction and institutions. We have to accept the need to justify philosophy's role and function in contemporary society, and to show what makes the study of philosophy worthwhile—but we should do so without reducing the subject matter at hand.

So why should you embark on the study of philosophy? What is it that makes this endeavor a great investment in your life's prospects? What is learned and conveyed by its study that both explains the subsequent success in 'real life' of those with degrees in philosophy, and equally defines the deep excitement and satisfaction that every true student of philosophy readily testifies to?

The answer, I believe, consists in the philosophical focus on deep issues of thought and life that resonate with one's overall existence. Philosophy focuses on issues of dramatic importance, and even though philosophy addresses these ideas with regard to their validity in the medium, as it were, of pure thought and discourse, the issues and ideas remain always connected to your existence and its cultural and social contexts.

Studying philosophy thus finds its value neither in the focus on a future career nor is it a self-sufficient exercise in the ivory tower of academia; rather, its true value consists in the existential significance of its probing attitude defined by investigating and challenging the taken-for-granted meanings and beliefs that define our cultural and social worlds.

Philosophy is here defined as an activity; it is a reflexive approach towards underlying assumptions within one's own understanding and vis-à-vis the encountered phenomena. Philosophy as such an activity entails some formal and general elements which transcend particular traditions, approaches, or disciplines.

The significance of philosophy as a reflexive practice can be brought out by four of its most central aspects: (1) the historical and cultural importance of philosophical ideas; (2) critical thinking as a core feature of the cognitive processing of ideas and arguments within philosophy; (3) the explicit articulation of domain-specific core assumptions in cultural and social discourses; and (4) the normative dimension of philosophical dialogue as a utopian ideal for democratic society. In what follows I devote brief discussion to each to then sum up the value of philosophy and its study in a final paragraph.

The Impact of Philosophy

The cultural and historical impact of philosophical ideas is immense. It often goes unnoticed simply because the ideas have long been integrated into the semantic fabric of our cultural self-understanding. Philosophy as a reflexive practice articulates its concepts and ideals often in critical opposition to prevailing ill-founded thoughts and practices. Early Greek thought engaged in a critique of anthropomorphic thinking to establish basic principles of reality independently from projecting human attributes onto nature.

The subsequent development from Thales up to Democrites, Socrates and Plato, and Aristotle established philosophy as a practice of thought engaged in a double dialogue: philosophy came to relate reflectively and analytically to the prevailing cultural concepts and practices, as much as it also relates to its own philosophical perspectives, constantly self-critiquing and improving on previous conceptions. This double orientation towards the culture as such and towards its own tradition continues to define the uniquely embedded and yet critical nature of philosophy.

In this process, philosophical ideas become a shaping force with regard to cultural and historical movements, including scientific perspectives, religious traditions, as well as social and political movements. As much as the Greek critique of mythology was essential for an early scientific thought, the thought of Bacon and Descartes and their critique of ideology and tradition helped establish the path of the modern scientific worldview. Similarly, the French and American revolutions are inconceivable without 17th and 18th century Enlightenment thought.

The ideas of Hobbes, Locke, Hume, and Rousseau developed visions of government, republic, and society that crushed the old and gave birth to new political orders. G.W.F. Hegel's political thought mirrors the ideals of self-government and mediates the republic with individual rights and civil society, but it is Hegel's philosophy of history and its idea of necessary progress that shaped much of the 19th and 20th century through Marx and his followers.

It is not necessary to argue that philosophical ideas 'cause' these movements and perspectives purely by means of their ideas. But it is certainly plausible to reject a purely materialistic view that economic or otherwise social conditions alone explain their force. It seems that this force of conceptual articulation has its own symbolic power. Take a concept of the individual and inalienable human rights pertaining to it, and you see what a concept can do, what social and political force it could unleash.

To be sure, the study of those ideas in philosophy involves the danger of abstracting them from their contexts, to analyze them as pure ideas severed from culture and history. And yet, the opposite move to reduce those ideas as nothing but expressions of cultural and social contexts—as in a purely historical approach—would also miss the mark.

Instead, studying philosophy enables you as the student to engage in the critical interpretation, analysis and assessment of the ideas. It takes the ideas face value in their claim to say something true and meaningful, and to engage this claim to see whether it holds up! In this vein, we would still understand Descartes' aim for a first philosophy, for an unshakable foundation of all thought, in its historical light stemming from the decline of medieval certainties (triggered by events like the discovery of the New World, Kepler, Copernicus, and Galileo, the national Bible translations and invention of the printing press).

Yet the validity of Descartes' ideas would not be judged relative to the context, but based on assessing their evidential and argumentative force, which is done, to be sure, influenced by our own background assumptions. Accordingly, the study of philosophy attempts to mediate taking the context into consideration and yet enabling you to critically participate in an assessment of the issues at stake.

I would like to highlight three very recent developments in philosophical thought that explicitly establish a relation between the validity of ideas and their cultural and social contexts:

Applied ethics emerged from the perceived need to bring ethical theory to bear on the concrete moral dilemmas as they emerge in cultural and professional situations. Often highly specialized pertaining to the fields of health care, business, environment, etc., the new sub-fields like bioethics, business ethics, environmental ethics, and others mediate the reflexive insight and analytic rigor of philosophical thought with the case-by-case scenarios of agents situated in professional fields.

The increasing awareness of globalization led to a decentralized view of philosophy itself, allowing Chinese philosophy, African philosophy, Islamic philosophy, for example, to emerge as serious perspectives within philosophy itself. Similarly, concepts like cosmopolitanism, international law, and intercultural dialogue responded to the new globalized world order, in which philosophy participates in the development of a new globalized self-understanding.

Finally, we witness new critical philosophies that change our ideas about the grounds of self, society, and culture. Critical social theory, post-structuralism, feminist philosophy and neo-Marxism developed radical critiques of basic assumptions pertaining to the subject, the idea of progress in history, the autonomy of scientific truth, the nature and reach of power, as well as deconstructing essentialist conceptions of sex as well as gender, the biological essentialism of race, and unchecked normative ideals of culture as emergent from class and social status. These new ideas have been very influential in the current humanities and social sciences.

Critical Thinking

It is a well-known and well-grounded fact that the study of philosophy is highly conducive to acquiring critical thinking skills. In a sense, philosophy is critical thinking in pure form. Philosophy's almost complete loss of a true and unique domain of experience, due to the constantly increasing conquest of the understanding of reality through the natural and human sciences, further enforces its focus on the construction and interpretation of arguments. Logic as a heuristic device for the representation of impeccable inferences lies at the core of the formal aspect of argumentation and shapes, with its rigor of formal coherence, every other aspect and domain of philosophical inquiry.

Students should take at least one course in logic to assure an understanding of how formal inferential chains are to be constructed. To be sure, the focus of philosophy still remains a substantive understanding of underlying premises and assumptions to advance the understanding of issues. What is at stake in philosophical inquiry, after the demise of the last great systems (in German idealism) that made claims as to integrate all actual knowledge within their philosophical frameworks, is to get at what is necessarily presupposed but not yet articulated in explicit understanding.

Since the articulation of basic concepts and methodological or ontological presuppositions is the true object of philosophy, its attentiveness to the very construction of arguments lies naturally at its core. A philosophical argument is one in which the premises are fully laid bare, the inferences towards a conclusion follow with stringency and are sound, and the pros and cons that could be brought forward with regard to the defended conclusion are included in the analysis.

It seems at first preposterous that philosophy claims to be the central discipline with regard to critical thinking skills. And not to be misunderstood, critical thinking is no doubt present and in many ways taught in any academic discipline. Yet what defines philosophy uniquely with regard to critical thinking comes into focus when we pay closer attention how and what is learned in philosophy courses. Those courses are generally related to some sub-discipline, like epistemology, metaphysics, ethics or ethical theory, philosophy of mind or language, for instance; importantly, the subject matter is not approached by merely setting up some problem or issue to proceed towards its discussion and analysis; nor are empirical opinions towards those domains, say of scientists or laypersons, collected and discussed.

Instead, philosophy approaches its subject matter via the mediation of major thinkers and their approaches. Philosophy thus actualizes its rich tradition by looking at the major discursive expressions vis-à-vis a particular issue. The basic point is that the access to the philosophical issues is through the understanding, analysis, and critique of classical philosophical texts. In other words, philosophy deals first and foremost with arguments, and the arguments are found in complex texts; they are expressed by philosophical authors from whose texts they have to be culled so as to advance the understanding of the issue.

The lifeblood of philosophy is therefore the systematic interpretation of texts with regard to their inherent claims about a subject matter, such that the prevailing argument is first reconstructed so as to in turn construct an argument that may validate or improve upon the position represented, undertaken with the aim to more deeply and correctly understand the subject matter at stake. It is clear how this approach differs from the history of ideas since it makes the arguments systematically as strong and coherent as possible in order to assess them with regard to their validity.

Furthermore, it becomes clear that this focus on distilling the argumentative substantive from the realm of diverse textual, cultural, and historical forms of discourse trains you as the interpreter to address the cognitive core of a position—thus to unfold the structure of critical thinking as found in the text, to subsequently subject it to the critical thinking as practiced by you as the interpreting student.

The interpretation of texts is done for the sake of the mining of new thoughts, challenging visions different perspectives, and opposing viewpoints the reconstruction of which defines the core dimension of the academic study of philosophy. And the reconstruction of the arguments encountered in the text is, as we saw, always guided by the epistemic interest of assessing their validity. Philosophical interpretation is essentially a practice of reflective judgment, which involves an orientation towards the soundness of a position while retaining the sense of context and cultures that stands behind the stated positions and that provide experiential and evidential resources for the expressed views. Critical thinking in philosophy is thus never a mere means, it is never a pure skill practiced for its own sake, but always oriented towards the better and advanced comprehension of the subject matter at stake.

As such, it elevates and embeds the formal coherence provided by logic into the higher aim of understanding the underlying assumptions and concepts of the phenomena under investigation so as to better understand the issues themselves. In a word, in philosophy critical thinking is both the object and the subject of the inquiry, since its object, the texts, express critical thinking about a subject matter, and its own analysis exemplifies critical thinking so as to assess its validity.

Philosophy as Reflexive Practice

We saw that philosophical ideas have considerable impact in history and culture, and that philosophy is uniquely equipped to convey critical thinking skills. Yet these features still miss an additional major benefit of the study of philosophy which can be brought out by emphasizing the essential connection between skill and content, between cognitive approach and domain of experience.

Philosophical reflection is the reflexive appropriation of the underlying beliefs and assumptions that define domains of experience. The philosophical analysis of these basic concepts or frameworks fulfils a double function with regard to the agent and her social world. By reconstructing the ideas or concepts that underlie a certain social domain of action, philosophical reflection both distances the agent from the immediate immersion into the context, while it also deepens her involvement in that context by reconstructing its value or purpose.

For example, a philosophy course in the "Philosophy of Art" creates a reflexive distance towards the artworld and art history by analyzing the structure of aesthetic experience, the function and value of art, and relevant concepts such as beauty and authenticity. Yet by doing so it similarly deepens and intensifies the appreciation of artworks and artistic movements as such, and reconnects and embeds the experience of art with the agent's lifeworld and her cultural existence.

The idea of this reflexive concept of philosophy consists in utilizing and unleashing the critical acumen of philosophical thought with regard to the cultural and social domains within which we as agents are usually situated. Philosophy is thus not conceived as a self-sufficient activity separated from our everyday existence, discovering a realm of objects or concepts that exists apart or above other things. Rather, philosophy is like a consciously acquired shadow, a critical mirror of what happens, of how we exist and interact socially, and as such it helps to create both a distancing and deepening relation to the taken-fort-grated meanings and practices that shape our self-understanding.

Immanuel Kant famously defined the essential questions of humankind as "What can I know?" "What should I do?" and "What may I hope for?" but was quick to emphasize that all three culminate in our quest to understand "What is the human being? (Was ist der Mensch?)" Now the first two questions still stake out major areas of philosophical inquiry (the theory of knowledge (epistemology) and moral philosophy (or ethics), yet I suggest that it is more productive to take our cue by looking at the question regarding the human being.

If we now understand the human being not as defined by an "essence" and "nature" conceived as an eternally fixed entity, but understand it as a being that realizes itself in culture and society, we can redefine philosophical activity as the reflexive understanding of human existence via the reconstruction of its major domains of self-expression.

Philosophy thus becomes the reconstruction of those domains of experience within which we as human generally live and act. It addresses the major issue about ourselves, about what the human being is, by analytically approaching the underlying assumptions in the realms of cultural objectifications and practices.

Equipped with this look, philosophy ceases to be discipline that has a few central areas, like metaphysics and epistemology (being and knowledge), moral philosophy (ethics), and logic, together with a mushrooming field of so many sub-disciplines and divisions. Rather, the whole richness of topics and courses in philosophy can now be understood as contributing to our self-understanding as we exist in the culture and society.

Specifically, while still honoring the important role of epistemology and ethics, we may now see metaphysics together with philosophy of religion as oriented towards the ontological foundations of our being; aesthetics and philosophy of art as addressing the sensuous dimension of our embodied experience; political philosophy as addressing our political structures, struggles, and ideals; philosophy of history as appropriating the unique dimension of history; the philosophy of science as reconstructing how science is possible, and the philosophy of social science as understanding the understanding of cultural agency and social institutions. Many sub-divisions like philosophy of music (aesthetics) and philosophy of technology (politics) would fit into these, while others, like the philosophy of language or philosophy of mind, would seem to address more basic or general aspects of knowledge and reality.

The crucial point is not to establish a new order of those domains which exist as clusters of experience rather than in terms of a foundational hierarchy, but rather to emphasize the special relations between a particular philosophical practice of problematization and a specific field of experience, as for instance between the philosophy of education and education as a cultural practice. Philosophical discourse establishes itself as a reflexive reconstruction of the domain-specific semantics which constitute our social and cultural discourses.

You as the student of philosophy thereby become a more knowledgeable, thoughtful, focused, and informed participant in those experiential domains. In this vein, the benefit of the study of philosophy consists in an advanced capacity to creatively participate in the ongoing reproduction of those social and cultural domains, which is due to the critical distance that philosophy creates with regard to their underlying concepts and values.

Philosophical Dialogue as Political Ideal

Reconstructing the underlying conceptual frameworks and value-orientations of our cultural practices enables the student to get a critical distance that is constructively directed towards a better engagement with the diverse domains of experience. The benefit of philosophy here consists in enabling a richer and fuller participation in social practices. This leads directly to our last major point, the role that philosophical education can play for our political community. Being a productive agent who constructively contributes to the diverse fields of science, education, art, law, etc., equipped with a deeper and reflexive understanding of their grounds, needs to be accompanied by a general sense of being a citizen in a diversified and decentralized society.

Apart from the functional fields of action, our society defines itself as a democratic one in which agents respect one another as equally worthy of recognition, as deserving the same respect and the same treatment regardless of their different social and cultural backgrounds and aspirations. The commitment to a moral framework of mutual respect and recognition normatively grounds, like an ethical umbrella that covers the diverse fields of discourses and practices, all social action. As we saw, philosophical inquiry reconstructs basic concepts of those social discourses and cultural practices, but the way philosophical discourse engages these problematics entails another source of social utility, namely the intersubjective respect of the other.

Philosophical inquiry, which we saw to be mediated by the argumentative reading of texts, presupposes that the other can follow and appreciate a rational reconstruction of the issues at stake. It entails such respect in its dialogical orientation towards the other, in which dialogue is understood as the truth-conducive back-and-forth between my own and the other's view on a shared issue, the engagement of which reveals limits as well as insights in both of our views.

There is an implicit ethic at work here, a normative infrastructure of presupposing that the other and oneself are both committed to their respective views yet also rational enough to let go if a better view emerges, if the other is right. All this is oriented towards a better grasp of the issue, and yet equally aware that any such view will again be mediated by one's views, that it emerged from the situation.

Philosophical dialogue can, if understood in this vein, serve as a model for a public sphere and political community that orients itself in its procedures at democratic intuitions. The normative core of this idea of dialogue goes back a long way, as it defines the Socratic critique of the sophists. The discursive capacity to engage an issue in the dialectical pro and con so as to advance the understanding is here opposed to the reductive use of this process in an instrumental fashion, as a strategy that merely advances one's social power by means of rhetorical force. Instead, the power of productive dialogue is now seen as standing in the service of a new shared vision with transforms all involved and thus serves the common good.

The study of philosophy teaches this approach, and it does so in the most effective manner: by simply doing it, by entailing and practicing it. A dialogical ethos of respecting the other by taking her claims seriously is what defines philosophical activity. This is not a theory that is taught, but a practical orientation which undergirds all truly philosophical inquiry.

Because it constitutes the embedded normativity of philosophical inquiry, it is able to create a practical disposition, a discursively quasi-embodied capability in the students of philosophy. It is nevertheless important to articulate this underlying ethos, in part to guide the student and in part to remind her of the value of openness, especially with regard to unusual or different perspectives within and outside of philosophical discourse.

In our context, articulating such an ethos helps bring out the major politico-normative benefit of the study of philosophy. Philosophy courses devoted to ethical theory, applied ethics, political philosophy, democratic theory, global ethics, etc. are to play a central role in this mission.

The relation between philosophical study and civic membership opens up a huge issue not only for philosophy itself, but for higher education and its purpose in general. Politicians too often fall prey to reducing college to the production of job-conferring degrees, instead of seeing its function in the wider context of producing agents capable of participating productively in society as such, including the democratic political sphere. Philosophy is here part of rediscovering the role that humanities and social sciences are to play in this context. Its own focus, besides the ethos of dialogical inquiry, is naturally to be found in ethics and social and political philosophy.

Philosophers have played an active role in redefining our sense of democracy, as their contributions to democratic theory, global ethics, and cosmopolitanism demonstrate. Figures such as John Rawls, Jürgen Habermas, Martha Nussbaum, Peter Singer, and Richard Rorty, to name a few, have become prominent voices within and outside of academic philosophy. They were able to bridge between philosophical discourse and the democratic public. Politics itself, however, realized the need to refocus our attention, as the US Department of Education's commissioned study "A Crucible Moment: College Learning and Democratic Future" attests to. The Association of American Colleges & Universities heads this call with strong and multifold support, and the increasing number of centers and institutes devoted to civic discourse and engagement at American universities is promising.

Philosophy courses devoted to ethical theory, applied ethics, political philosophy, democratic theory, global ethics, etc. are to play a central to in this mission. On the one hand focused on normative core concepts and basic value-orientations, and on the other hand aiming at an analysis of the social and political conditions, these courses mediate the 'abstract' philosophical thought with non-ideal social and political reality. They prove indispensable for anyone eager to act politically in light of a moral ideal, without closing one's eyes to the conditions of its realizability.

To sum up: Our essay presented you with the idea of philosophy as a cognitive activity embedded in the lives of agents—agents who engage their core assumptions reflexively by articulating them in the course of philosophical study. Thus, the crux of this idea of philosophy is the reflexive mediation of cultural traditions as well as social and political interactions as maintained by philosophically educated agents. As shown, philosophical education entails a set of features, tied together by the interest in understanding the underlying assumptions and premises of certain fields of experience and action, and coupled with a keen sense of the individual agent's own existence.

Philosophical activity combines the rational reconstruction of domains of experience, which get thereby articulated in specific modes of thought, with the reflexive self-understanding that such reconstructions are always partial, emergent from contexts, and shaped by cultural and social backgrounds. Yet in philosophy, the agent's context-dependency is not objectified as an empirical aspect of the self, but rather seen as an unavoidable and productive backdrop for experiential understanding.

The aim remains always to reach a better, more adequate and coherent understanding of the issues at stake. Philosophical inquiry is thus defined by the reflexive awareness of one's ongoing and necessarily incomplete self-understanding, but its quest is the most accurate and satisfying account of the phenomena themselves. Its reflexive practice aims to reconcile a certain epistemic humility—one which entails an ethos of openness vis-à-vis new experiences and different perspectives—with its commitment to scientific rigor and analytic clarity. In short, studying philosophy:

provides you with an in-depth insight into the conceptual and value-theoretical foundations of political, religious, and scientific perspectives, engaging you to reconstruct the bases of the claims to truth and rightness that inhere in these human projects;
conveys to you a set of critical thinking and communication skills which are honed by reconstructing complex arguments presented in philosophical texts
engages you to understand and appreciate the core concepts and values presupposed in the diverse cultural discourses in different fields of society, and
equips you with a dialogical ethos of respect that provides both a practical disposition as well as a utopian ideal for political action within a democratic society.

That's what studying philosophy is good for. Now who is still surprised that philosophy students land on their feet when looking for a job?

Chapter 22: Why Should Students Choose Public Speaking as their Undergraduate or Graduate Major/Minor?) - Professor Joseph Burns, PhD

Joseph Burns, PhD
Professor of Communication
Southeastern Louisiana University

Joseph E. Burns, Ph.D. is an Associate Professor of Communication at Southeastern Louisiana University. An 11-year radio veteran, his teaching areas include Public Speaking, digital audio, broadcast news, copywriting, film history, criticism and ethics. His current research topics include rock music, and radio broadcasting. When not working Dr. Burns is an avid woodworker and wood turner and performs a weekly syndicated rock trivia show, Rock School, on the Southeastern radio station.

The Truth about Public Speaking

Public Speaking has always been a tough sell when it comes to a choice for a major or minor in higher education.

I believe the main reason is that, to those who decry it, it sounds like something anyone is able to do. We can all talk. Just get up in front of a group of people and, well, talk. There. Public Speaking.

At it's base denotative definition, that's true. Taking that logic we can all sing, play the drums and play the violin as well. Everyone of us can open our mouths and strike a tone, beat a skin with a mallet and draw a bow across four strings.

My guess is that you've been a room with someone who's tried any one of those three without being schooled first in the skill.

Painful, wasn't it?

Public Speaking, like singing, playing the drums and playing the violin is a skill. None of us are born with the ability to do it well. It is a set of abilities and patterns as well and a performance aspect that allows a single person to communicate to an audience in such a way that what is said is not only entertaining but is remembered and later recalled.

That's Public Speaking.

"But," you say, "I know people who are just born public speakers. They always seem to know what to say. They're good in front of a crowd. They're clever. They're witty. They're just good at it."
We all know these people, right? They're just good at it. It is seemingly born into them. Here's the truth. It's not. What you're seeing, when you see these good speakers, is someone who has from a very early age, been practicing their craft. Now, I don't doubt the person is an extrovert, but make no mistake, What you're seeing is the product of years of practice and probably a lot failure.

That person, from an early age, was trying out lines, facial expressions, laughs, comments, body language, etc. Without knowing the academic terms, he or she was practicing Public Speaking.

That may sound like just a sales pitch but let me contrast it. I'll bet you've also been to the party where there's been the person who was also practicing but wasn't quite catching the non-verbal cues from those around him or her. This is the loud, brash person who is telling the joke that's a little too off-color. He or she is a little too loud, a little too forward and a little taken aback when people leave because of something he or she said.

That's Public Speaking as well. It's just not very refined.

Why Study Public Speaking?

Are you beginning to see why you should study it? Let's go on.

Too often people place too much emphasis on the "speaking" side of Public Speaking. I've been teaching the craft now for close to two decades and one of the mistakes I see is, "If I just get the words right, that's enough."

Personally, I believe it stems from the academic world's emphasis on English and writing. Watch your spelling. Grammar is of the utmost importance. Do not split your infinitives.

A student will prepare a "speech" and step to the podium to deliver it but once he or she begins to speak, they're all alone. Either they're reading to the audience or they're so involved in the text that performance and body language is non-existent. Don't get me started on the monotone voice.

When I don't hand them an A+, they're shocked. "But what I said was perfect. It was grammatically correct and the research was top shelf!"

Probably. To be honest I can't remember a thing you said, and I was being paid to listen to you. Imagine how little the students in class were paying attention.

There's an old adage I like to pull out early in my classes that goes, "Your audience will judge you 70% on how you look, 20% on how you sound and 10% on what you say."

Are those percentages dead on? I doubt it, but they make the point you need to know. What you say is nothing without the performance to back it up. You must be able to stand before an audience with confidence, dressed appropriately, able to deliver your text with authority.

"If you're excited, I'm excited"

"If you're excited, I'm excited."

It's true. I've listened to speeches on topics I could not have cared less about but the speaker delivered the topic with such love and admiration, I couldn't help but care.

You should be able to raise and drop your voice to create tension. You should be able to move your person away from the podium and use not only your arms but your entire body to drive a point home. You should be able to use a simple voice inflection to make a group of people know that what you said was so important that they should write it down.

It truly is not what you say but how you say it. No one is born with that ability. It is a skill and it must be learned, practiced and perfected.

Are you beginning to see what you should study Public Speaking? Let's go on.

It is a stereotype, yes, but over my many years as a Public Speaking professor, I have found it to be true more often than not that those who study the arts are more receptive to learning Public Speaking than those who study the sciences or business. If you pit a Theater or an English major against a Physics or a Finance major, for the most part, the students of the arts will be far more open to the performance aspects of the skill.

To be fair, the students of the sciences almost always have better outlines and get their homework in on time.

So what?

I have never believed that the purpose of university study is to get a job. If you want to go to school after high school to specifically get a job, you should go to a vocational school. University or college is to make you a well-rounded individual schooled in multiple fields, hire-able because of your ability to think rather than because of your have a single specific skill.

Google What Employers are Screaming for

That said, at just about any job, in just about any field, employers are screaming for new employees with "communication skills."

Go ahead, look it up. Just hit the web and Google what employers are looking for. The results will avalanche out of your computer screen and at the top of the heap will be communication skills.

I did just such a search and I found the following terms: Communication skills, speaking skills, interpersonal skills, the ability to communicate effectively, verbal skills, communicate clearly and concisely, communicate wisely.

This skill is severely lacking in the working world today, clearly. It is what employers want.

I could ask once again at this point if you see why taking Public Speaking is viable but it would be a rhetorical question.

It is, however, the answer to a question: "How do you prove to an employee that you have communication skills?"

Think about that. How do you "prove" you have communication skills? The easiest way I know is to have a major or a minor in Public Speaking. That way you can show a possible employer your skill set and what you are able to do for his or her company communication wise.

Right now there is a serious lack of employees who can communicate. There must be otherwise there wouldn't be such a call for those who could. As a new employee who was willing to be the face and voice of a new product line, a group's presentations, even just the person who presents to the company itself, you would be invaluable.

As I often say to my science and business students, if you are that person who not only understands the science, understands the business model, but is willing to stand before a group and communicate it effectively, you'll be the go-to person of the building.

That is not at all understating the effect a solid public speaker can have on a business.

Let's go on.

It would seem from what I've written above that Public Speaking is an ancillary study. You have a major and you use Public Speaking as way to bolster what you do. Not so.

If you Use your Voice as your Main Tool, a Degree in Public Speaking is a Strong Choice

Majoring is Public Speaking is also a strong choice for an academic career, but what can you do with it?

Well, how about speak? I know that sounds like too simple an answer but think through all the professions that require you to speak. Have you ever thought of broadcasting, news anchor, voice-over artist, radio professional, training, or motivational speaking?

If you're a person of religion, Pastor, Preacher, Father, Minister.

In any occupation where you use your voice as your main tool, a degree in Public Speaking is a strong choice.

What about sales? A good salesperson must be able to read and understand the verbal and nonverbal responses of whomever he or she is talking with. Those are base skills taught as part of the Public Speaking major.

If politics have every interested you, Public Speaking would be a strong choice. Politicians make a great many speeches. In fact, it's their number one form of connecting to the voters who vote them into office.

As a corollary to a career in politics, a Public Speaking major would make a strong speech writer.

I could continue to list and list but it would probably be better for you to look at the lists yourself. To write the segment just above I went to three major job search websites and simply entered, “Public Speaking.” There were thousands of results, most listing “Public Speaking Skills” specifically.

“What are you most Afraid of? - "I'd rather Die”

Go ahead. Look for yourself. One more thing.

"I'd rather die.”

Have you heard the statement that statics show that people are more afraid of Public Speaking than death?

Yeah, me too.

It comes from the 1977, "Book of Lists" where they claim to have polled over 3000 people and asked their biggest fears. Public Speaking topped the list at 41 percent.

So "Death" had to be second, right? Well, no. Death was actually 7th behind heights, insects, financial problems, deep water and sickness.

A few polls since then have apparently supported the Public Speaking over death findings but please, if you were given the true choice of delivering a speech or jumping off the edge of the Grand Canyon to your certain demise, which would you do?

It's a set of false finding based of asking people, off the top of their heads, "What are you most afraid of?"

"This is How you Fail Public Speaking"

But still, I have students who seem deathly, sorry for the pun, afraid of Public Speaking. I have been given notes from parents, counselors, therapists, doctors and psychologists, all claiming some student has some disorder that should excuse him or her from delivering a speech.

Nope. I ignore them all. This is a Public Speaking course. You must public speak. You cannot get through chemistry without working with chemicals. You cannot get through Public Speaking without Public Speaking.

Interestingly enough, when you ask a person why they are so afraid of speaking, you rarely receive a definitive answer. They're not faking it. They're truly frightened. They just don't know why.

Furthermore, that fear can manifest itself if physical ways. I've had students vomit, cry, shake and simply refuse to leave their seat. I've had a student sit down on the floor behind the podium and another faint hitting her head on the way down opening up a rather large gash on the side of her head.

The best display of nerves came when a student stood at the podium, motionless and quiet for a solid minute before announcing, “This is how you fail Public Speaking,” and walking out of the room.

He never returned which is too bad. It was so funny I would have given him a C grade just for the entertainment value.

It all stems, I believe and I've yet to be proved wrong, from the fear of failure and judgment.

If you make a mistake, people will hear that mistake and they will laugh at you, mocking you.

This rarely happens. Audiences seldom want you to fail and furthermore, don't care or rarely ever notice your mistakes until you call attention to them.

Yet, the fear persists.

Is Public Speaking for you? Think Back

So, if you're nervous about Public Speaking, the major or minor is not for you, right?

That couldn't be farther from the truth. Think back. Were you ever nervous about something you do now on a regular basis? The first time you were pitched to in baseball. The first time you were with your boyfriend or girlfriend. He first time you flew on an airplane.

You were frightened then as well but you got past it. You overcame the fear and what you do has become second nature.

It's the same with Public Speaking. No one is born with the ability. It is a skill that must be learned and mastered just like a sport or a musical instrument. The difference is that you start this skill in front of a group of people.

Think back over what I have written thus far. The benefits of having this skill so far outweighs the short time it will take you to overcome the fear, and you will overcome the fear.

All of those student who had notes from their professional nay-sayers all stood up in my class and spoke. They all passed.

You will too.

Fears are to be faced, especially when the rewards are so great.

There are so many ways to defeat your fears in terms of Public Speaking. I don't personally recommend the now cliché imagining everyone in their underwear but if it works for you, imagine away.

Let's wrap it up.

Should you explore Public Speaking as a major or a minor. Well, I am certainly bias but my answer is an unequivocal, "Yes."

There is no other degree that spans so many fields and gives you so much benefit from a single skill set.

It is well worth your time to discuss the option with someone there at your university or college. The ability to express yourself is a skill that you'll never regret delving into.

As I said at the beginning of this piece, everyone can talk.

Not everyone can "speak."

Chapter 23: Why Study Political Science? – Professor John Dunn, FBA

John Dunn, FBA
Professorial Fellow
Cambridge University

*Professor John Dunn was born in Britain during the second year of the Second World War and grew up there, in Occupied Germany after the war, and in Iran and India. He has studied many aspects of the history of political thinking across the world and done research on contemporary politics in Ghana. He studied history at King's College, Cambridge from 1959 to 1964, and at Harvard from 1964-6, and has taught at King's from 1966 up to today. He was Professor of Political Theory at Cambridge University from 1987-2007, and has been a Visiting Professor at the Universities of Ghana, British Columbia, Bombay, Minnesota, Yale (three times), and Chiba in Japan. He has published over 20 books which between them have been translated into more than a dozen languages. He is best known for his view of how to understand the history of political thinking across time and space and recognize its strategic importance for recognizing what politics is, why it assumes the shapes which it does, and why it matters so overwhelmingly for human life. His most widely read books have been **The Political Thought of John Locke** (1969), **Modern Revolutions** (1972), **Western Political Theory in the Face of the Future** (1979), **Political Obligation in its Historical Context** (1980), **The Politics of Socialism** (1984), **Interpreting Political Responsibility** (1990), **The Cunning of Unreason** (2000), **Setting the People Free** (2005), and **Breaking Democracy's Spell** (2014).*

The Best Reason to Study Politics

The best reason to study politics is that it will be what decides the kind of future human beings are going to have for many centuries to come. To study it you must study why human beings find it so hard to live graciously and generously with each other beyond their own families, the arrangements they have so far thought of to handle that task for them, and why those arrangements have worked so poorly. You must explore why we are finding the task ever harder, as we need to live with each other on ever larger scales and with people about whom we know and understand less and less.

Over time politics has done us far more conspicuous harm than clear good. It is more urgent than ever to tilt that balance of harm and benefit back the other way. The only sound way to try to do so is to sharpen our understanding of why the balance comes out the way it does and raise our capacity to judge and choose together how to find and keep to a better way.

What Politics Really is, How it Works, and Why it Comes out as it Does

I have written three books over the last twenty five years which try to define this challenge. The first, *The Cunning of Unreason: making sense of polit*ics (HarperCollins & Basic Books 2000), began as an effort to show what politics really is, how it works, and why it comes out as it does. It addresses these questions at first at a very general level and shows why they are hard to answer.

It shows how to answer them through a single lengthy historical example: what happened to the politics of my own country, Britain, through the impact of one very drastic leader, Margaret Thatcher. It explains why she could have such an impact and why the impact she had was so far from what she had hoped. It is a complicated book by the end, and it would probably be best read last of the three.

Setting the People Free, and Breaking Democracy's Spell

The other two, *Setting the People Free: the story of democracy* (Atlantic 2005) and *Breaking Democracy's Spell* (Yale University Press 2014) are shorter and they move much faster. *Setting the People Free* shows what has happened in a remarkable change in the world's political imagination over the last two and a half centuries.

Three hundred years ago, virtually everywhere on earth, almost anyone who held beliefs on the question presumed that human beings were and must be ruled by single and very special persons over whom they plainly had no real control. That was what they expected, and insofar as they thought about it they mostly saw nothing wrong with it. It is not clear how far the practical realities have changed since, but today the great bulk of those who believe anything about the question believe something quite opposite.

They see it as the right basis for humans to govern one another for most of them to authorize personally in some way or other those who do the governing. They see the authority to provide that authorization as an entitlement of their own, and they see it as imposing definite responsibilities on their governments (the duty to serve them effectively and in ways they endorse). They also now in many settings see it, a bit trustingly, as imposing clear limits not just on their government's right to act for them and in their name, but also on its power to do so.

Setting the People Free describes the main way in which the change has come about, through the choice and diffusion of the ancient Greek word *demokratia* (the rule of the people) as the name for legitimate government across the world. It shows what an odd choice in many ways that was, how late in the word's history the choice was made, and why it has proved so deeply confusing. It shows the name and idea of democracy caught up along that pathway in three dramatic political arenas.

The first was in ancient Athens, which gave the word to the world as a serious category of political understanding and offered an astonishingly vivid and creative collective life to try to understand through it.

The second, over two thousand years later, was the political birth of the United States, still the most powerful state in today's world. Since that time America's thinkers, politicians and citizens have come to confuse their own remarkable political arrangements with an ideal political framework for the rest of mankind, with consequences you will all have noticed.

The third was France's great and terrible revolution, which spelled the long term historical doom of monarchy as a real form of government and brought out by its bitter close why democracy will always remain such a deeply unstable and deceptive basis for understanding political reality.

What makes it so is its offer of political equality to all citizens. It was in the closing stages of France's revolution that the unreality of that offer first came into clear focus. What makes it unreal is two tough and rigid features of the way the world really is. The first is the huge gap in power between rulers and ruled in every state. The second is the relentless dynamic of economic inequality in the way a global capitalist economy works. That is what the politics of today's world is mainly about. We have been exploring just what it means ever since.

Breaking Democracy's Spell shows how these confusions work in the politics of today's world. It shows why democracy has come to carry such different meanings in America, or Britain, China or India, and across the other continents of the world. It shows why Americans or Britons disagree so bitterly with most of their Chinese or Russian contemporaries, and with devout Egyptians, Taliban militants or dedicated Al-Qaeda fighters, about what democracy is or should be, and what claims it has on anyone's respect and loyalty. It shows how badly our smug assumption that our way of governing ourselves is simply the right way has distorted our recognition of the ugliest, most foolish and most dangerous of the ways in which our societies are now living.

You too can Think about What the Politics of your World Means

You too can think about what the politics of your world means, and what you and others can do about it by watching or reading the news in any medium, and talking to your friends and enemies as you move through the rest of your lives. If you care at all about the harm we are all doing to the world in which we have to live by the way we have so far chosen to do so, and still go on choosing every day, you will have a vast amount to think about for decades to come. You will think about it much better, and have a far better chance of seeing how to tilt that balance back in a less grim direction, if you spend some of your time while you are still relatively free in learning what really is going on in politics and why it is happening like that.

That is the opportunity which studying political science should offer you. The political scientists of the world must make sure, if you choose to take it up, that that is the opportunity you get.

Chapter 24: Using Humor To Help Students Get Comfortable With Political Science And With You, The Professor – *By Thomas Sutton, Ph.D.*

Thomas C. Sutton, Ph.D.
Department of Politics & Global Citizenship
Baldwin Wallace University, Ohio U.S.A.

Dr. Thomas Sutton directs the Community Research Institute (CRI), which engages students in applied research with local and regional nonprofit and government agencies. He coordinates the public & nonprofit management major/minor and the Africana studies minor. He also develops and coordinates innovation and entrepreneurship programs for the Center for Innovation and Growth (CIG).

Dr. Sutton serves as independent political analyst for WEWS Channel 5, providing periodic analysis of local and national political developments and events. Dr. Sutton also has 13 years of prior experience in nonprofit management, including work in civic education, services for homeless persons, and programming for juvenile offenders and at-risk youth.

How Humor Can Help Your Teaching

What makes for an effective class? Ultimately, it's about what students learn and experience. Ideally, students learn content and skills that will stay with them after the end of a course. Establishing a positive relationship with the class helps to get students to engage. All of us know that if we care about something, we pay attention. More importantly, we put in the effort to achieve whatever we seek from what we care about. If you care about your dog, you take the time to provide food, water, and walk the dog. If you care about your neighbors, or at least local city ordinances, you also pick up the poop (living on a main street in a city with more dogs than people, I really appreciate the 'picking up the poop' part of caring for a dog.

And no, I don't have a dog; just two turtles with a tank filter that takes care of the poop problem). If students care about the class for whatever reason (interested in the material, like the professor, seeking a good grade, avoiding parental retribution for poor performance), they are likely to put in the effort needed to do whatever the professor requires (excluding personal favors like running errands or babysitting your kids).

So how do you get students to care about your course? Develop a positive relationship with the class. This involves two elements. First, create a class environment that is inviting and enjoyable (dare I say, fun?) Using humor gets students' attention and helps them relax. The second element is building a relationship with students. Humor helps, but more important is demonstrating that you care about the students. This means that you want them to do as well. This does NOT mean getting involved in their personal lives (Danger, Will Robinson, Danger!)

For smaller classes (30 or less), use introductory activities the first week that ask students' names, home communities, major/minor, and something interesting (e.g., favorite movie/music genre), as well as what they want to learn about in the class. The key to success is to then learn the students' names over the next few weeks. Calling students by name makes a big difference. They know that you know them.

For a larger class where name memorization is impractical, using humor can help to establish general student interest. The focus in a large lecture class is on you as the instructor; ergo, you have the stage and can use humor to get their attention and help keep them focused on the material (and less interested in TikTok videos). Whether a large or a small class, it's just a lot more fun being occasionally humorous than being deadpan serious 24/7 (and if you are deadpan serious all the time, you're probably not reading this anyway).

Teaching and Risk-Taking

Effective teaching requires taking risks. Teaching new content, using technology (hyflex teaching, anyone?), and using discussion and group activity strategies all entail risk. Keeping a good sense of humor helps to ease the risk. If students trust that you generally know what you are doing and admit when failure happens (yes, dear colleagues, professorial pedagogical failure is possible), they will cut you a bit of slack. Being able to laugh at yourself helps a lot! Don't take yourself too seriously. Be willing to expose yourself (no, not that way...you do want to keep your job and avoid prison). We can all improve our teaching through practice, which necessarily means that the first try at a new technique will not be the best but can improve with practice.

Given the polarized political environment in which we are working, humor can help to take the edge off conflict. Our discipline seeks to understand political processes and dynamics. Politicians are the same as teachers; to be effective, they need to connect with their audience. Rather than trying to be the teaching comedian, just relate examples of effective use of humor by politicians and humorists.

Relevant to our current national spending spree is the Ronald Reagan quote: 'I am not worried about the deficit. It is big enough to take care of itself.' (The Independent 01/22/2009). We teach students about important, serious material, such as why presidential election polling is always accurate (Greenberg 11/04/2020), and that 3.14 out of 5 political scientists recommend representative democracy as a way to improve health, economic well-being, and life-span (no guarantees about whiter teeth). Our field includes stories of drama, tragedy, heroism, and romance (remember South Carolina Governor Mark Sanford and his 'hike' with a beautiful journalist along the southernmost part of the Appalachian Trail...in Argentina? Mrs. Sanford had a really hard time finding that trail on a map). It's important for students to learn theories and approaches to understanding politics, government, and international relations.

As professors, we walk the balance beam of being serious about teaching and learning, but need to remember not to take ourselves or our subject too seriously (yes, all students should take at least three courses in political science as the ultimately important field in academia, but...)

So Much Material, So Little Time!

There's a simple reason why politicians are the target of every late-night talk show host. Their personas and behavior get magnified and distorted by media coverage. And, what politicians and institutions do (or not do) have significant effects on average folks (just ask anyone about what they did with their Covid stimulus payments). A simple way to include humor in your teaching is to borrow from the best. Use quick clips of late-night routines in a visual presentation (thank goodness for YouTube – I'm never up late enough to watch these shows). Quote a line or two from a comedian that relates to the topic of the day.

I often teach on Friday afternoons, which requires a bit of levity to minimize student squirming (yes, just like waiting for the 9th period bell in high school). I end the class with five minutes of 'monologue of the week,' choosing a clip from one of the late-night hosts, who lately have had a hard time letting go of Trump as a target (Lawler, 04/23/2021). After showing a clip, I'll ask students if they had seen it before class, and how it relates to the material we covered that week. The kicker: bonus points on a quiz or test. Some students actually take notes!

Using humor in class does not require developing jokes. Truth is stranger, and funnier, than fiction (although these days, truth is contestable). Of course, another easy way to use humor is the classic: political cartoons. Inserting an occasional political cartoon into a PowerPoint can either illustrate a concept or help break up the complexity of material being discussed (Dougherty 2002). Political cartoons also represent a cultural approach to communicating events and political perspectives to a mass audience. According to Hans Speier, the use of wit and humor by the powerful and the people is found across many societies and epochs (Speier 1975).

There are so many absurdities in politics: QAnon, anyone? (Roose 3/4/2021). Politicians are public figures facing scrutiny that has been elevated by social media. And yet...they still do the dumbest things that most of us would never even consider (I hope). Exhibit A: former New York congressman Anthony Weiner (pronounced Weener, just like the...) tweeting pictures of his 'reproductive organ' to a woman, leading to his resignation from Congress in 2011. But that's not all: he did it again, as a candidate for Mayor of New York City in 2013, sending similar pictures to a 15-year-old girl, while sitting in bed with his young son. But don't cry for Weiner: after a year in prison for conviction as a sexual offender, Weiner is now CEO of IceStone, a company that makes countertops from broken glass (Siff 10/19/2020).

Tiptoeing Through the Minefields

Humor is a great approach for engaging students in class. However, there are topics that should be avoided. We are living in a culture with increased awareness of the need to recognize and respect groups that have historically been marginalized or at the very least, considered 'fair game' as comedic targets. It's okay to target public figures, provided they are known due to their choice of employment or pursuit. It's not okay to target people who become publicly known through no fault of their own. Discussing the roots and effects of the 'culture wars' in the U.S. and elsewhere is important, particularly in our discipline. Becoming a participant in the culture wars is not necessary and does not help students understand the divides between voices seeking to respect and protect particular groups and those who criticize this as 'political correctness.'

So, who and what are off limits in classroom humor? We used to say that this was similar to political campaigns of the last 20+ years (do not directly target an opponent's gender, race, or religion). Sadly, Donald Trump's presidential campaigns and sojourn in the White House buried the bar of public decency (somewhere between the sewer and gas lines). He succeeded at getting elected despite a slash and burn approach to all opponents, use of demeaning nicknames, embrace of negative stereotypes that demeaned women, minorities, and people with disabilities, and a platform defining immigration from anywhere except Norway as a national threat. It is challenging but important to show how and why these tactics were used and are tempting for candidates and officeholders to embrace, but that cause much more harm than good to a country's democratic norms, social relations, and public image.
The point is simple: do not use Trump as an example of how to treat students (or anyone). Do not mock a person's gender, race, ethnicity, or religion. Do not joke about physical and developmental disabilities, sexual identity, gender identity, and body size. Joking about horrific events such as genocide, famine, and war are off limits. Avoid off-hand remarks trading in stereotypes about urban neighborhoods or poor rural regions of the country (particularly since you may have students with family from these types of places). Regarding appropriate language (e.g., cussing): keep it clean! Do not channel your inner Lenny Bruce/Richard Pryor in a classroom. An occasional curse may slip out – recognize, apologize, and move on.

Most important of all: do not ever, ever joke about, make fun of, or mock a student. No matter how dense or disconnected they may appear to be, Do. Not. Mock! Even a gentle off-hand comment can be taken the wrong way. Mocking is like lying. Your reputation as a good professor who cares about their students, like an honest reputation sullied by a lie, can be damaged quickly if you are perceived to be making fun of a student.

Even with these cautions in mind, it's likely you will make a mistake, as I have several times. In one instance during the first week of class, students were introducing themselves. A quiet Black student with a long African-sounding name and a slight accent introduced himself. I jokingly responded, 'now that's quite an Irish name!' The student responded quietly, 'Yes, I am Irish.' And I laughed. Bad move. The student was born and raised in Dublin, Ireland, to parents who had immigrated from Nigeria in the 1980s. He dropped the class; I never saw him again. Does this mean never try to be funny in class? Of course not. It does mean two things: avoid personal jokes about students and be ready to apologize for the times when mistakes are made. We are all imperfect humans.

But I'm not Funny – Should I Try?

Using humor in a classroom can be a challenge for professors who do not consider themselves to be funny. It is important to be credible. 'Fake it 'til you make it' can result in students questioning your authenticity, which can get in the way of developing a positive, productive classroom environment.

Be willing to share a bit about yourself within whatever you define as your bounds of comfort. Punctuating a presentation with a brief story about yourself, a family experience, being a parent, can provide a bit of levity and help students connect with you as an individual. As figures of authority (hopefully, respected), with power over students' futures, we balance the importance of credibility through expertise with the value of connecting with students to develop an effective teaching environment.

As your experience teaching and getting to know students deepens, you will find topics and techniques that some students will find humorous. For instance, generational cultural differences (e.g., not getting their cultural references) can often produce a few giggles. As a father of now twenty-something daughters, I get a lot of mileage out of 'bad Dad jokes.' The language and references of social media change rapidly, which typically results in using terms and phrases that students think are no longer 'cool.' Using memorized jokes can also work when the joke fits the topic being taught. I use a goofy joke to test students' understanding of the mean average in an introductory statistics course:

A lawyer, a doctor, and a statistician go bowhunting for deer. They walk into a field where an impressive buck with a full antler rack is standing about 100 feet away. The doctor aims and releases, and the arrow lands 50 feet behind the deer. The lawyer takes aim, releases, and the arrow lands 50 feet in front of the deer. The statistician gets very excited and shouts, 'we got him, we got him!'

While eliciting a few groans, jokes like this show that you have the confidence to look a bit silly without losing credibility. Again, this technique is only effective if you are comfortable using this approach!

Another approach is to try using material from the 'political humor pros.' Jon Stewart was so focused on getting his audience to understand U.S. politics and government that he wrote a mock textbook, 'America: The Book,' (Stewart, Karlin, and Javerbaum 2004). A number of political science faculty used the book as part of introductory U.S. government courses. While a popular trend for some faculty, one experimental study indicated that use of the text did not have a positive effect on test scores, but that use of humor in teaching is more interesting and engaging for students (Baumgartner, Morris 2008).

Oh, This Is Easy – I'm Naturally Funny! (Whoa, Partner...)

A paradox involved in using humor in the classroom is that people who are naturally funny may need to be more careful about their use of humor than folks who do not have this type of talent. Teaching about politics and government involves the occupational hazard of cynicism, which in the hands of faculty with a sense of humor, can either produce a lot of laughs or the need for apologies. It's tempting to throw acerbic darts at politicians and government agencies, but too much of this results in feeding students with negative stereotypes. Balance and moderation, grasshopper; balance and moderation...

People who have a knack for spontaneous humor need to continuously screen their reactions. A student may say something that invites a snarky response (like the student walking into a final exam for U.S. Government and Politics asking, 'Professor, what are the three branches again?') The mantra that should hang like a banner in your cerebral cortex is 'Filters, baby, filters – not just for cigarettes anymore!' Save the reaction for happy hour with friends and colleagues (but even there, be careful not to mention students by name). It should also go without saying that what is funny at happy hour may not be appropriate in the classroom.

The 'filter function' becomes even more challenging when discussing sensitive topics for which you have strong personal opinions. A discussion about whether state governments should restrict transgender students to sports involving their birth sex will elicit much stronger opinions and reactions than sharing your research on European Union trade policy.

While comedians may have a field day (yes, pun intended) with this, as a professor, you need to approach the topic with your best 'objective, neutral, explain and listen to all sides' teaching persona. You may personally think abortion is murder or that restricting women's reproductive rights is medieval, but the topic needs to be addressed in a balanced manner, without resorting to humor to 'lighten things up.'

Remember the minefield? It's not just about avoiding offending students. It's also about understanding that a careless comment or crude joke could land you in the Title IX (U.S. sexual harassment law) investigator's office. Legally, a Title IX complaint relies on the perception of the accuser using the rational person standard to determine if a comment or action can be considered an instance of sexual or racial harassment. Which means, 'But I was only joking' is not a defense.

As a former sexual harassment investigator at our university, this was the most common form of complaint from students. Faculty think they are just being funny, but if what you say is perceived as demeaning and offensive, you could be liable. However, this does not mean avoiding sensitive topics such as gender and racial disparities, religious conviction as a basis for political engagement, or how demographic identities influence public opinion. If a topic and how it is treated is justified as part of the discipline, you are on safe ground. Just be sure to disable the humor switch while navigating the minefield.

If It Isn't Enjoyable, Why Are You Teaching?

Let's face it: teaching is a challenge. Each day in a classroom is a performance, trying to capture the attention of the audience and hope that they get your message. Using humor as a teaching technique helps to get students' attention and to develop a sense of connection between you and the class, and ideally to build their interest in the course content. Using humorous video clips, political cartoons, goofy jokes, and sharing personal stories are all ways in which you can build these connections and help students learn. The best part? Teaching becomes a lot more fun for you!

References

Baumgartner, J. C. and Morris, J.S. (2008). 'Jon Stewart Comes to Class: The Learning Effects of America (The Book) in Introduction to American Government Courses.' *Journal of political science education*, 4:169–186.

Dougherty, B. K. (2002). 'Using Political Cartoons in the Classroom.' *International Studies Perspectives*, August 2002, Vol.3, No.3, 258-270.

Greenberg, D. (2020, November 4). 'The Political Scientist Who Warned Us About Polls.' *Politico*. https://www.politico.com/news/magazine/2020/11/04/decades-ago-one-political-scientist-rejected-political-polling-as-faulty-and-futile-maybe-the-world-should-have-listened-434106.

Lawler, K. (2021, April 23). 'From Fallon to 'SNL' to Oliver, how late-night TV is moving on (or not) from Trump.' *USA Today*. https://www.usatoday.com/in-depth/entertainment/tv/2021/04/23/late-night-tv-how-fallon-snl-moving-or-not-trump/7157491002/.

Roose, K. (2021, March 4). 'What is QAnon, the Viral Pro-Trump Conspiracy Theory?' *The New York Times*. https://www.nytimes.com/article/what-is-qanon.html.

Siff, A. (2020, October 19). 'What is Anthony Weiner Doing Now?' NBCNews, New York. https://www.nbcnewyork.com/news/politics/iteam-election-project/4-years-later-after-prison-stint-anthony-weiner-reflects-on-possible-role-in-2016-election/2676817/).

Speier, H. 1975. Wit and Politics: An Essay on Laughter and Power. *American Journal of Sociology*, Vol. 103, No. 5 (March 1998), 1352-1401.

The Independent. (2009, January 22). 'Ronald Reagan.' https://www.independent.co.uk/news/presidents/ronald-reagan-1482923.html.

About Dr. Thomas Sutton

Dr. Thomas Sutton directs the Community Research Institute (CRI), which engages students in applied research with local and regional nonprofit and government agencies. He coordinates the public & nonprofit management major/minor and the Africana studies minor. He also develops and coordinates innovation and entrepreneurship programs for the Center for Innovation and Growth (CIG).

Dr. Sutton serves as independent political analyst for WEWS Channel 5, providing periodic analysis of local and national political developments and events. Dr. Sutton also has 13 years of prior experience in nonprofit management, including work in civic education, services for homeless persons, and programming for juvenile offenders and at-risk youth.

---

[i] Sigmund Freud, *Civilization and Its Discontents,* trans and ed. James Strachey (New York: W.W. Norton, 1961), 49.

[ii] A term used by John Locke in the *Second Treatise of Government,* chapter two, and first theorized by Thomas Hobbes in *Leviathan,* chapter 13,14.

[iii] John Locke, *The Second Treatise of Government* (New York: MacMillan, 1952), 9.

[iv] Emma Goldman, " Anarchism: What It Really Stands For," in *The Broadview Anthology of Social and Political Thought*, vol 2, ed. Andrew Bailey, et. al. (Ontario: Broadview Press, 2008), 19.

[v] Commanitarianism is associated with the political writings of, among others, Alasdair MacIntyre, see, *After Virtue (*Notre Dame, Indiana: University of Notre Dame

Press, 1981) and Michael J. Sandel, *Liberalism and the Limit of Justice* (Cambridge: Cambridge University Press, 1982). In brief, proponents of commanitarian thought criticize liberalism for promoting a lack of commitment to common shared values.

[vi] Jean Jacques Rousseau, *The Social Contract and Discourses,* trans. G.D.H. Cole (London. Everyman's Library, 1973) chaps, VI-VIII, bk.1.

[vii] Aristotle, *The Politics,* trans. T.A. Sinclair (London: Penguin Books, 1992) 183-84.

[viii] MacIntyre, *After Virtue,* especially, chaps. 14-16.

[ix] Harold Lasswell, *Politics: Who Gets What, When and How* (New York: McGraw-Hill, 1936).

[x] 496 U.S. 310(1990).

[xi] David Easton, *The Political System* (New York: Knopf, 1953).

[xii] Michael Sandel, "The Procedural Republic and the Unencumbered Self" in *The Broadview Anthology of Social and Political Thought,* Essential Readings, ed. Andrew Bailey, et al (Ontario: Broadview Press, 2012) 944-52.

[xiii] Thomas Hobbes, *Leviathan,* in *The Broadview Anthology,* Essential Readings, 298.

[xiv] Aristotle, *Politics,* 452.

[xv] The various forms of government is not discussed in this chapter but most basic political science books are more likely to discuss them. See for example, Lowi et al. (2013).

[xvi] See Harold Lasswell (1936). *Politics: Who gets what, when, and how*. New York. McGraw-Hill.

[xvii] Here the author is defining politics from a more authoritative aspect of government: See David Easton (1971). *The Political system, 2nd ed.* New York. Knopf.

[xviii] The author in this definition focused on the human activities, which need coordination from a recognized structure. See Karl Deutsch (1963). *The nerves of governmen*t. New York: Free Press.

[xix] This definition maintains that politics cannot exist independent of the community, which means there must be a given society or community depending on how that society is defined for politics to occur. See Dernard Crick (1972). In defense of politics, 2nd ed. Chicago: University of Chicago Press.

[xx] This definition was adopted from APSA website. See http://www.apsanet.org/content_9181.cfm?navID=727

[xxi] The various subfields are provided at http://www.apsanet.org/sections/index.cfm?CFID=28712062&CFTOKEN=72438920. It is possible that other areas of study could be added to the list provided here.

[xxii] James Madison, "Letter to William T. Barry, August 4, 1822," in *Writings,* ed. Jack N. Rakove (New York: Library of America, 1999), 790.

[xxiii] Some of my publications included: Antwi-Boasiako, Kwame Badu, William Forbes, Ben Dixon (2014). Some Fundamentals of Conservation in South and West Africa. *Environmental Ethics.* 36: (1) 5-30. (http://www.pdcnet.org/pdc/bvdb.nsf/purchase?openform&fp=enviroethics&id=enviroethics_2014_0036_0001_0005_0030) DOI: 10.5840/enviroethics20143613. The co-authors were from different departments.

See also Antwi-Boasiako, Kwame Badu and Peter Csanyi (2014). "Introduction: Decentralization from global perspective." In *The theories of Decentralization and local government: Implementation, Implications, and Realities: A global Perspective*. Edited by K.B. Antwi-Boasiako and Peter Csanyi, pp. 9-24. Stephen F. Austin State University Press, Nacogdoches, Texas. Here the co-author is from Europe.

[xxiv] See also the "*Future of Political Science: 100 Perspectives*" edited by Gary King, Kay Lehman Schlozman, and Norman H. Nie in 2009. Published by Routledge. This edited book presents various views and sometimes opposing views on the future of the discipline.

[xxv]Robert Dahl, *Who Governs?* New Haven: Yale University Press, pp. 223-226.

[xxvi] Jeb Barnes, Dust-Up: Asbestos Litigation and the Failure of Commonsense Policy Reform Georgetown: Georgetown University Press; see also Jeb Barnes and Thomas F. Burke, How Policy Shapes Politics: Courts, Rights, and Litigation and the Struggle over Injury Compensation New York: Oxford University Press 2015.

[xxvii] Geoffrey Tweedale, Magic Mineral to Killer Dust: Turner & Newall and the Asbestos Hazard, New York: Oxford University Press 2000.

[xxviii] Robert A. Kagan, Adversarial Legalism: The American Way of Law, Cambridge, MA: Harvard University Press; see also Andrea Boggio, Compensating Asbestos Victims: Law and the Dark Side of Industrialization Surrey, UK: Ashgate 2011.

[xxix] Dennis Cauchon, "The Asbestos Epidemic: An Emerging Catastrophe," *USA Today*, February 8, 1999.

[xxx] Barnes, *Dust-Up* at Chapter 4.

[xxxi] Ibid. at 33-42.

Made in United States
North Haven, CT
04 November 2026

21767035R00108